Advances in Molecular Oncology

ADVANCES IN EXPERIMENTAL MEDICINE AND BIOLOGY

Editorial Board:

NATHAN BACK, *State University of New York at Buffalo*
IRUN R. COHEN, *The Weizmann Institute of Science*
ABEL LAJTHA, *N.S. Kline Institute for Psychiatric Research*
JOHN D. LAMBRIS, *University of Pennsylvania*
RODOLFO PAOLETTI, *University of Milan*

Recent Volumes in this Series

Volume 596
MECHANISMS OF LYMPHOCYTE ACTIVATION AND IMMUNE REGULATION XI
Edited by Sudhir Gupta, Frederick Alt, Max Cooper, Fritz Melchers and Klaus Rajewsky

Volume 597
TNF RECEPTOR ASSOCIATED FACTORS (TRAFs)
Edited by Hao Wu

Volume 598
INNATE IMMUNITY
Edited by John D. Lambris

Volume 599
OXYGEN TRANSPORT TO TISSUE XXVIII
Edited by David Maguire, Duane F. Bruley and David K. Harrison

Volume 600
SEMAPHORINS: RECEPTOR AND INTRACELLULAR SIGNALING MECHANISMS
Edited by R. Jeroen Pasterkamp

Volume 601
IMMUNE MEDIATED DISEASES: FROM THEORY TO THERAPY
Edited by Michael R. Shurin

Volume 602
OSTEOIMMUNOLOGY: INTERACTIONS OF THE IMMUNE AND SKELETAL SYSTEMS
Edited by Yongwon Choi

Volume 603
THE GENUS YERSINIA: FROM GENOMICS TO FUNCTION
Edited by Robert D. Perry and Jacqueline D. Fetherson

Volume 604
ADVANCES IN MOLECULAR ONCOLOGY
Edited by Fabrizio d'Adda di Fagagna, Susanna Chiocca, Fraser McBlane and Ugo Cavallaro

A Continuation Order Plan is available for this series. A continuation order will bring delivery of each new volume immediately upon publication. Volumes are billed only upon actual shipment. For further information please contact the publisher.

Fabrizio d'Adda di Fagagna
Susanna Chiocca
Fraser McBlane
Ugo Cavallaro
Editors

Advances in Molecular Oncology

Edited under the auspices of the European Institute of Oncology (IEO) and The FIRC Institute of Molecular Oncology Foundation (IFOM)

 Springer

Fabrizio d'Adda di Fagagna
FIRC Institute of Molecular Oncology
 (IFOM)
20139 Milan
Italy
fabrizio.dadda@ifom-ieo-campus.it

Susanna Chiocca
European Institute of Oncology
 (IEO)
20139 Milan
Italy
susanna.chiocca@ifom-ieo-campus.it

Fraser McBlane
European Institute of Oncology
 (IEO)
20139 Milan
Italy
fraser.mcblane@ifom-ieo-campus.it

Ugo Cavallaro
FIRC Institute of Molecular Oncology
 (IFOM)
20139 Milan
Italy
ugo.cavallaro@ifom-ieo-campus.it

Library of Congress Control Number: 2007921525

ISBN-13: 978-0-387-69114-5   e-ISBN-13: 978-0-387-69116-9

Printed on acid-free paper.

© 2007 Springer Science+Business Media, LLC
All rights reserved. This work may not be translated or copied in whole or in part without the written permission of the publisher (Springer Science+Business Media, LLC, 233 Spring Street, New York, NY 10013, USA), except for brief excerpts in connection with reviews or scholarly analysis. Use in connection with any form of information storage and retrieval, electronic adaptation, computer software, or by similar or dissimilar methodology now known or hereafter developed is forbidden.
The use in this publication of trade names, trademarks, service marks, and similar terms, even if they are not identified as such, is not to be taken as an expression of opinion as to whether or not they are subject to proprietary rights.

9 8 7 6 5 4 3 2 1

springer.com

Contents

Foreword vii

Section 1. Genome Instability, Checkpoints, and the Cell Cycle   1

1. Comparison of DNA Replication in *Xenopus laevis* and *Simian Virus 40* 3
   Maren Oehlmann, Cathal Mahon, and Heinz-Peter Nasheuer

2. A Genetic Screen Implicates miRNA-372 and miRNA-373 as Oncogenes in Testicular Germ Cell Tumors 17
   P. Mathijs Voorhoeve, Carlos le Sage, Mariette Schrier, Ad J.M. Gillis, Hans Stoop, Remco Nagel, Ying-Poi Liu, Josyanne van Duijse, Jarno Drost, Alexander Griekspoor, Eitan Zlotorynski, Norikazu Yabuta, Gabriella De Vita, Hiroshi Nojima, Leendert H.J. Looijenga, and Reuven Agami

3. An Investigation into 53BP1 Complex Formation 47
   Kevin C. Roche and Noel F. Lowndes

4. Short Abstracts – Session I 58

Section 2. Tumor and Microenvironment Interactions   65

5. Tumor Promotion by Tumor-Associated Macrophages 67
   Chiara Porta, Biswas Subhra Kumar, Paola Larghi, Luca Rubino, Alessandra Mancino and Antonio Sica

6. The AP-2α Transcription Factor Regulates Tumor Cell Migration and Apoptosis 87
   Francesca Orso, Michela Fassetta, Elisa Penna, Alessandra Solero, Katia De Filippo, Piero Sismondi, Michele De Bortoli and Daniela Taverna

7. Modulatory Actions of Neuropeptide Y on Prostate
   Cancer Growth: Role of MAP Kinase/ERK ½ Activation 96
   *Massimiliano Ruscica, Elena Dozio, Marcella Motta,
   and Paolo Magni*

8. Short Abstracts – Session II 101

Section 3. Animal Models   105

9. The *Arf* Tumor Suppressor in Acute Leukemias:
   Insights from Mouse Models of Bcr–Abl-Induced
   Acute Lymphoblastic Leukemia 107
   Richard T. Williams and Charles J. Sherr

10. Short Abstracts – Session III 115

Section 4. Transcription and Epigenetics   119

11. Short Abstracts – Session IV............................ 121

Section 5. High-Throughput Approaches and Imaging   127

12. Identification and Validation of the Anaplastic Large
    Cell Lymphoma Signature 129
    Roberto Piva, Elisa Pellegrino, and Giorgio Inghirami

13. Cell-Cycle Inhibitor Profiling by High-Content Analysis 137
    Fabio Gasparri, Antonella Ciavolella, and Arturo Galvani

14. Short Abstracts – Session V 149

Section 6. Novel Pathways and Therapeutic Targets   153

15. Regulation for Nuclear Targeting of the Abl Tyrosine
    Kinase in Response to DNA Damage.................... 155
    Kiyotsugu Yoshida

16. Short Abstracts – Session VI........................... 166

Section 7. Poster Abstracts   171

Author Index....................................... 255

Subject Index....................................... 265

Foreword

Recent advances in molecular oncology have shown us the importance of genes in tumor formation and growth. The second International IFOM-IEO Meeting on Cancer has focused on such currently relevant topics in this area as genome instability, novel technologies, transcriptional regulation mechanisms, and the identification of therapeutic targets. This meeting, promoted by the European School of Molecular Medicine (SEMM) and the University of Milan, in collaboration with IFOM (The FIRC Institute of Molecular Oncology Foundation) and IEO (European Institute of Oncology) involved important presentations by eminent cancer researchers from all over the world.

The major sponsor of the event was the Umberto Veronesi Foundation (FUV). With 200 participants, 34 oral presentations and 86 posters, the Meeting offered the possibility of a "full immersion" in the field, with presentations of the latest and most relevant findings in molecular oncology. The chapters in this volume are organized around the six scheduled sessions that took place during the meeting, and the abstracts and expanded abstracts fall into these six different topics: Chapter 1: Genome Instability and Mechanisms of Cell Cycle Control; Chapter 2: The Influence of Microenvironment on Cancer (interactions between cancer cells and their surrounding tissues); Chapter 3: Animal Models; Chapter 4: Cancer Epigenetics (functional alterations in the genome which contribute to tumor onset); Chapter 5: Imaging Systems and High-Throughput Technologies (technologies that allow the rapid and simultaneous analysis of tens of thousands of genes and proteins); and Chapter 6: Identification of Therapeutic Targets. Topics were chosen to emphasize the importance of translational research.

Among the attending speakers was David M. Sabatini, from the Whitehead Institute in Cambridge (USA), who set up a specific gene platform allowing the selective inactivation of single genes within an organism. This technique is now leading to the understanding of what happens when a particular gene is unable to work properly. Alberto Mantovani, from the Istituto Clinico Humanitas (Rozzano) and the University of Milan presented his studies linking cancer and inflammation; Carlo M. Croce, from the Comprehensive Cancer Center (Ohio State University, USA), referred to his recent analysis of the role of micro RNAs in tumorigenesis. RNA (ribonucleic acid) is best known for being a molecule that contains "transcribed" information from within the DNA, that is then "translated" into proteins. Carlo Croce spoke about so-called micro-RNAs, unusual tiny bits of RNA that regulate the expression of genes potentially involved in tumor formation. William C. Hahn, from the Dana-Farber Cancer Institute in Boston (USA), in a collaborative effort with David Sabatini, has been investigating the role of oncogenes and tumor suppressor genes, and has created a number of tumor model systems with a well-defined genetic background. This approach is proving to be a critical tool to understand the genetics of cancer. Giulio F. Draetta from the Merck Research Laboratories in Boston (USA), Pier Paolo Di Fiore, IFOM Scientific Director, Olivera J. Finn, from the University of Pittsburgh (USA), and David M. Livingston from the Dana Farber Cancer Institute in Boston (USA) presented research dedicated to the identification of new therapeutic targets and intelligent drug design. Dr. Finn, in particular, spoke about tumor vaccines and immunotherapy. Dr. Livingston, who is also a member of the Advisory Board at IEO, gave the closing lecture, presenting recent results obtained in breast and ovarian tumors.

The members of the Scientific and Organising Committee of the Second International IFOM-IEO Campus Meeting on Cancer were: Fabrizio d'Adda di Fagagna, Director of the IFOM research programme "Telomeres and cellular senescence", Ugo Cavallaro, Director of the IFOM research programme "Cell adhesion and signalling in tumor progression and angiogenesis", Susanna Chiocca, Director of the IEO research programme "Viral control of cellular pathways and biology of tumorigenesis"; and Fraser McBlane, Director of the IEO research programme "Molecular mechanisms of leukemogenesis".

Section 1
Genome Instability, Checkpoints, and the Cell Cycle

Chapter 1
Comparison of DNA Replication in *Xenopus laevis* and *Simian Virus 40*

Maren Oehlmann, Cathal Mahon* and Heinz-Peter Nasheuer[†]

Department of Biochemistry, National University of Ireland, Galway, Ireland,
maren.oehlmann@nuigalway.ie,
*cathal.mahon@nuigalway.ie,
[†]h.Nasheuer@nuigalway.ie

Abstract. DNA replication is a fundamental process within the cell cycle. The exact duplication of the genetic information ensures genome stability. Extensive research has identified the principal players required for the sequential processes: origin-licensing (a controlled order of events giving a chromosome site the potential to be initiated within the S phase of the same cell cycle); initiation (by removing the license a previous licensed site is transformed into a site where the DNA helix starts to melt); and DNA replication (copying the parental DNA by leading and lagging strand DNA-synthesis). The present report compares the advantages and limitations of studying DNA replication in the model systems *Xenopus laevis* (*X. laevis*) and in *Simian Virus 40* (*SV40*).

1 DNA Replication of *SV40*

With the exception of the hyperthermophilic archaeon *Sulfolobus solfataricus* (Robinson et al. 2004) and *Sulfolobus acidocaldarius* (Lundgren et al. 2004) all prokaryotes and viruses so far contain one origin. The *SV40* origin as a specific sequence was described by Bergsma (Bergsma et al. 1982).

The genetic information of *SV40* is encoded in a small (5,243 base pairs), circular genome, which encodes only six viral proteins. Three of the proteins are the capsid structural proteins (VP1, VP2, and VP3) and the other three are regulatory proteins (large T-antigen, small t-antigen and agnoprotein). To replicate its double-stranded DNA the virus mainly depends, as is the case with most other viruses, on its host cell. Besides the virus-encoded large T-antigen all proteins necessary for the replication of *SV40* originates from primate cells. To avail of the host replication proteins *SV40* has to drive the cells into S phase, a process achieved by interaction of the large T-antigen with the tumor suppressor proteins pRb and p53, and other factors. Moreover, *SV40* DNA replication depends on S phase cyclin-dependent protein kinases (Cdks) (Voitenleitner et al. 1999).

The 64 bp core origin of *SV40* is characterized by four pentanucleotides (GAGGC), an early palindrome (EP) sequence and an AT-rich tract. The *SV40* large T-antigen recognizes the *SV40* origin and forms a double-hexamer (Joo et al. 1998). The formation of the double hexamer initiates with the binding of a monomer to the pentanucleotide 1 of the origin. Further monomers associate and the "flanking" EP sequence supports the formation of the first hexamer (Kim et al. 1999). Pentanucleotide 3 is required for the assembly of the second hexamer. The origin-bound double hexamer introduces two structural changes within the origin: the melting of 8 bp within the EP sequence and the unwinding of the AT-rich sequence (Borowiec and Hurwitz 1988; Parsons et al. 1991). This "structural distortion" is believed to occur independently of the helicase activity of the large T-antigen but the underlying mechanisms are not yet understood (Simmons et al. 2004). The formation of the initiation complex further requires three host cell proteins: replication protein A (RPA) (Collins and Kelly 1991; Pestryakov et al. 2003; Weisshart et al. 2004), DNA polymerase α/primase (Murakami et al. 1986; Matsumoto et al. 1990; Dornreiter et al. 1993; Weisshart et al. 2000) and topoisomerase I (topo I) (Trowbridge et al. 1999; Gai et al. 2000; Weisshart et al. 2000). RPA is essential to bind to unwound single-stranded DNA (ssDNA) to prevent its re-annealing. DNA polymerase α/primase is required to synthesize RNA primers and their initial elongation whereas topoisomerase I relieves torsional strain ahead of the fork.

After the initiation of DNA synthesis, its elongation depends on further proteins such as DNA polymerase δ (Garg and Burgers 2005; Pavlov et al. 2006), proliferating cell nuclear antigen (PCNA) (Oku et al. 1998) and replication factor C (RF-C) (Bylund et al. 2006).

Due to the 5′ to 3′ activity of DNA polymerases it is fundamental for the DNA replication machinery to coordinate the synthesis of the continuous leading strand with the discontinuous lagging strand (Hubscher et al. 2000). Furthermore, no known DNA polymerase is able to synthesize DNA de novo. The synthesis of the lagging strand requires the synthesis of the Okazaki fragments each starting from an RNA primer, followed by Okazaki fragment maturation. Okazaki maturation factors are for example Flap endonuclease 1 (Fen1) and DNA ligase I (Waga et al. 1994). Fen1 takes part in removing the RNA primers, whilst DNA ligase I joins the gap between adjacent Okazaki fragments. Fen1 and DNA ligase I are recruited to replication sites by PCNA (Leonhardt et al. 1998; Warbrick 2000). Dna2 and RNase H are additional factor in the maturation of Okazaki fragments with Dna2 having helicase and exonuclease activity whereas RNase H specifically cleaves RNA hybridized to DNA (Kao and Bambara 2003).

2 DNA Replication in *Xenopus laevis*

In contrast to *SV40*, the genome of the eukaryote *X. laevis* is much larger (3.1×10^9 bp) and the genetic information is separated into 18 separate units – the chromosomes. In order to replicate the genetic information of each chromosome, in theory, at least one origin per chromosome would be required. Furthermore, in contrast to *SV40* replication, DNA synthesis in *X. laevis* is restricted to one per cell cycle and each synthesis phase is strictly followed by cell division. Experiments on mammalian chromosomes showed that this increased regulation of DNA synthesis requires each eukaryotic chromosome to initiate several origins to complete DNA synthesis within S phase (Huberman and Riggs 1968). Huberman and Riggs isolated and spread DNA from a cell culture, which for a few minutes had been incubated with ^{3}H. They noticed radioactive tracks spaced 50,000–200,000 bp apart. This method revealed not only multiple origins per chromosome but also bidirectional replication.

In the *X. laevis* early embryo, origin spacing on chromosomal DNA is roughly 5–15 kb (Blow et al. 2001). In preparation for DNA replication in S phase origins are licensed from late mitosis to early G1, a period of the cell cycle when Cdks are inactive (Nguyen et al. 2001). Licensing leads to the loading of the minichromosome maintenance proteins Mcm2 to Mcm7 (Mcm2–7) (Chong et al. 1995; Labib et al. 2000; Gillespie et al. 2001) and requires the ordered function of the origin recognition complex (ORC) (Burkhart et al. 1995; Lei et al. 1996; Rowles et al. 1996; Donovan et al. 1997; Mahbubani et al. 1997; Edwards et al. 2002), the cell division cycle protein 6 (Cdc6) (Hartwell 1973) and the Cdc10-dependent transcript 1 (Cdt1) (Maiorano et al. 2000; Tada et al. 2001; Devault et al. 2002; Tanaka and Diffley 2002). Geminin is an inhibitor of Cdt1 and prevents Mcm2–7 chromatin loading and is therefore a licensing inhibitor (McGarry and Kirschner 1998).

With the binding of Mcm2–7 a pre-replicative complex (preRC) is formed (Diffley 2004; Blow and Dutta 2005). At the G1/S transition, the preRC gets activated and thereby it is transformed into a pre-initiation complex (preIC). This requires the chromatin binding of *X. laevis* Cut5 and the action of the kinases Cdc7 and Cdk2 (Jares et al. 2000; Walter 2000; Van Hatten et al. 2002; Kubota et al. 2003). Furthermore, *X. laevis* Cdc45 and *X. laevis* GINS (Go, Ichi, Nii, and San for five, one, two, and three in Japanese; subunits: Sld5, partner of Sld five 1 – Psf1, Psf2, Psf3) chromatin loading are mutually dependent and required for initiation (Kubota et al. 2003; Forsburg 2004). After Cdc45 chromatin loading, DNA starts to unwind and RPA stabilizes the single-stranded DNA (Walter and Newport 2000). This is followed by the binding of DNA polymerases. As the DNA is replicated the license is removed and another license can only be acquired after passing through mitosis.

In *X. laevis*, ORC binds chromatin in a sequence-independent manner (Blow et al. 2001; Cvetic and Walter 2005). Chromatin saturation experiments with ORC showed that one ORC complex is bound every 8–15 kb DNA, closely matching the number of initiation events (Rowles et al. 1996). Interestingly, after licensing, Mcm2–7 complexes are approximately 10–20 times in excess compared to chromatin bound ORC (Mahbubani et al. 1997; Edwards et al. 2002; Oehlmann et al. 2004). It has previously been shown that ORC chromatin binding is stabilized in the presence of geminin (Oehlmann et al. 2004; Waga and Zembutsu 2006). Waga and Zembutsu (2006) showed that in the presence of geminin more than 10 ORC molecules bound to the plasmid was reduced to one or a few ORC complexes after licensing.

This result is in agreement with Rowles et al. (1996) showing a less tight ORC chromatin binding as licensing occurs – an observation previously termed as "licensing-dependent origin inactivation". Therefore, it might be possible that each ORC complex only loads one or two Mcm2–7 hexamers (Waga and Zembutsu 2006). It will be interesting to elucidate if before licensing occurs ORC binds chromatin in a cluster or in a spreaded manner. It is also important to note, that chromatin bound Mcm2–7 complexes and ORC do not always co-localize (Edwards et al. 2002), that Mcm2–7 proteins are only found on unreplicated stretches of DNA (Laskey and Madine 2003) and that initiation has been shown at sites in a distance from ORC where Mcm2–7 complexes are bound (Laskey and Madine 2003; Danis et al. 2004). One might therefore speculate that each chromatin site with a bound Mcm2–7 complex is a potential origin (Woodward et al. 2006). Interestingly, pulse-labeling experiments of newly synthesized DNA showed that initiation events do not occur equally spread over the genome but in clusters (Mills et al. 1989). Furthermore, even the chromatin binding in a ratio of 1 to 2 Mcm2–7 complexes per initiation event still supports the normal replication rate (Mahbubani et al. 1997; Edwards et al. 2002; Oehlmann et al. 2004) of 10 nucleotides per second per fork (Mahbubani et al. 1992). Therefore, the question arises what is the need for binding up to 10 times more Mcm2–7 complexes. This "MCM paradox" (Hyrien et al. 2003) might be explained by excess Mcm2–7 complexes licensing dormant origins, which are only used under replicative stress, allowing to complete DNA replication under low tread circumstances (Blow and Dutta 2005; Woodward et al. 2006).

Mcm2–7 complexes have multiple functions during DNA replication. They are required for the formation of the preRC, the formation of the preIC, origin unwinding and elongation (Forsburg 2004). In addition, Mcm2–7 proteins have a function in transcriptional activation, DNA damage response and chromatin remodeling (Bailis and Forsburg 2004; Forsburg 2004; Snyder et al. 2005). Sequence comparisons of Mcm2–7 proteins showed that all 6 Mcm proteins contain a highly conserved sequence covering approximately 200 nucleotides (Koonin 1993). Further investigations of this sequence revealed that Mcm proteins are a subgroup of the large AAA+ ATPase family, which also includes ORC subunits, CDC6 and RF-C (Iyer et al. 2004). Characteristic of proteins of the AAA+ ATPase family is the presence of the two conserved ATPase consensus motifs Walker A and Walker B. One of the possible functions of proteins of the AAA+ ATPase family is to act as a helicase (Neuwald et al. 1999; Iyer et al. 2004). All known helicases share three common characteristics. They are able: (a) to bind nucleic acids, (b) bind and hydrolyse NTPs, and (c) couple hydrolysis with the unwinding of the duplex nucleic acids in the 3′ to 5′ or 5′ to 3′ direction (Koonin 1993; Tuteja and Tuteja 2004). All DNA helicases therefore always contain a DNA-dependent NTPase activity (Koonin 1993). Indeed Mcm2–7 proteins have been shown to bind to DNA and are capable to bind and hydrolyse ATP (Lee and Hurwitz 2000). There are various data suggesting that all 6 Mcm proteins are essential and required for DNA synthesis (Labib et al. 2000). Furthermore, Mcm2–7 proteins predominantly form a heterohexamer in *X. laevis* (Kubota et al. 1997) and bind as a heterohexamer to chromatin (Prokhorova and Blow 2000). Interestingly, a recent study showed that ATPase activity of the Mcm2–7 complex is not required for preRC formation but is required for DNA unwinding. Until now, no helicase activity has been described for the Mcm2–7 complex. Nevertheless, in vitro helicase activity by the Mcm4/6/7 subcomplex was shown in different species (Ishimi 1997; Lee and Hurwitz 2000; You et al. 2002; Kaplan et al. 2003), but such a complex does not support DNA replication in Mcm depleted *X. laevis* extract.

One kinase required for the transition from preRC to preIC is Cdc7. This serine/threonine kinase Cdc7 is regulated by Dbf4 and Drf1. Recent studies showed that Dbf4 is expressed throughout development whereas the Drf1 protein seems to disappear after the mid-blastula transition (MBT), a developmental stage after which maternal messages are degraded and the early embryo starts to transcribe zygotic genes (Nakamura et al. 2000; Korner et al. 2003). In contrast to human and mouse cell lines no zygotic Drf1 transcript has been discovered in *X. laevis* (Montagnoli et al. 2002; Yoshizawa-Sugata et al. 2005). The egg extract contains a five-fold excess of Cdc7/Drf1 over Cdc7/Dbf4. Both kinase complexes have a comparable specific activity and it seems that Cdc7/Drf1, and not as previous reported Cdc7/Dbf4 (Costanzo et al. 2003; Yanow et al. 2003; Jares et al. 2004), is predominantly required for the activation of the preRC in egg extracts (Silva et al. 2006). Interestingly, even a Cdc7 depleted extract is capable of supporting partial replication (Takahashi and Walter 2005; Silva et al. 2006).

Recent studies also imply that origin activation in *X. laevis* is further regulated by histone acetylation (Danis et al. 2004; Iizuka et al. 2006). XHbo1, the *X. laevis* homolog of the human histone acetyltransferase HBoI, was shown by XHbo1 immunodepletion to be required for the acetylation of histone H4 before the preRC formation. Interestingly, human Hbo1 was able to acetylate Orc2, Mcm2, Cdc6, geminin and itself. Furthermore, XHbo1 coimmunoprecipitated geminin and Cdt1 and prevented chromatin binding of the Mcm2–7 complex and subsequent replication (Iizuka et al. 2006). This implies that Mcm2–7 chromatin binding is further regulated by histone and/or preRC acetylation. Furthermore, DNA methylation at CpG nucleotids prevents ORC chromatin binding and replication (Harvey and Newport 2003).

Models for the initiation of replication in *X. laevis* and *SV40* are shown in Figures 1 and 2, respectively.

3 Origins

From viruses to human, replication has to occur in order to ensure survival. The original replicon model proposed by Brenner, Cuzin and Jacobs, based on studies in *Escherichia coli* suggested the necessity for two components for

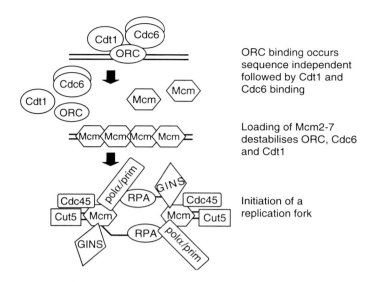

FIGURE 1. Model of the initiation of DNA replication in *X. laevis*. Origin recognition complex (ORC); cell division cycle protein 6 (Cdc6); Cdc10-dependent transcript 1 (Cdt1); Minichromosome maintenance proteins (Mcm); cell division cycle protein 45 (Cdc45); (Cut5); Sld5, Psf1, Psf2, Psf3 (GINS); replication protein A (RPA); DNA polymerase α/primase (polα/prim). In short: In order to prepare for replication initially ORC, Cdc6, and Cdt1 bind to chromatin and load Mcm2–7. Then with the help of Cdc45, GINS, Cut5, and other proteins origin activation occurs. It follows the loading of other proteins including RPA and DNA polymerase α/primase and DNA replication initiates.

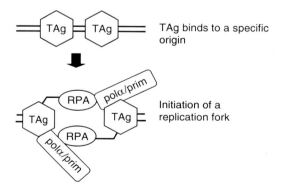

FIGURE 2. Model of the initiation of the *SV40* replication origin. Large T-antigen (TAg) replication protein A (RPA); DNA polymerase α/primase (polα/prim); the initiation of *SV40* DNA replication requires less factors than the cellular DNA replication since the multifunctional large T-antigen intrinsically carries these activities. That way, the virus avoids various levels of host regulation and is able to replicate more than once per host cell cycle.

the initiation of DNA synthesis, the replicator and the initiator (Jacob and Brenner 1963). The replicator was defined as a DNA sequence, which is recognized by an initiator protein. Replicator sequences have been identified in viruses and prokaryotes. The replicator sequence of *SV40* origin was described by Bergsma (Bergsma et al. 1982).

Great efforts were undertaken with the hope of identifying a universal eukaryotic replicator sequence. To date the origins of the eukaryote *Saccharomyces cerevisiae* (*S. cerevisiae*) are the best characterized. The *S. cerevisiae* origins were identified in transformation experiments and designated as autonomously replicating sequences (ARS). Some *S. cerevisiae* DNA sequences inserted into plasmids allowed this extrachromosomal DNA to replicate autonomously (Hsiao and Carbon 1979; Stinchcomb et al. 1979; Stinchcomb et al. 1980).

In contrast to the origins in *S. cerevisiae*, the replicators in higher eukaryotic organisms are poorly understood. The origins in higher eukaryotes lack consensus sequences and localize to intergenic regions. Moreover, the usage of origins has been shown to change with development (Maric et al. 2003). It has been proven in *X. laevis* that until the blastula stage of the embryonic development, any DNA can act as a replicator sequence (Harland and Laskey 1980; Hyrien et al. 1995). It is likely that epigenetic factors, as changes in the chromatin structure throughout development, influences origin usage (Gilbert et al. 1995; Danis et al. 2004). Until now no DNA sequence or structure common to several identified replicator sequences has been identified (Gilbert 2004).

4 Advantages and Limitations of Investigating DNA Replication in *Xenopus laevis* and *SV40*

Various key players of DNA replication have been initially determined in *X. laevis* and *SV40*. Therefore, both systems have been very successful in deciphering the process of DNA replication.

4.1 SV40 as a Model System – Advantages and Limitations

SV40 has evolved mechanisms enabling it to transport its genetic material into the nucleus of the host. *SV40* is capable of infecting mammalian cells, independently of cells dividing or resting. Once infected, *SV40* uses the DNA replication machinery of primate cells and therefore information about *SV40* DNA replication is of human relevance. Furthermore, the *SV40* genome is well characterized and its unique origin and sequence structures are known.

SV40 replication can be investigated in vivo and in vitro. It is possible to transfect primate cells (e.g., COS-1) with the virus and to replicate it. After extract preparation, only plasmids containing the specific *SV40* replication origin will replicate. It is further possible to replicate *SV40* origin-containing plasmids in extracts prepared from uninfected cells, if T-antigen is added. The use of the host replication machinery, with the exception of T-antigen, made *SV40* replication a powerful model to identify the proteins involved in human replication, downstream of replication-origin unwinding.

Since *SV40* manipulates the host cell to enter S phase, *SV40* is not a suitable model to investigate the regulation of eukaryotic origin initiation. Furthermore, the *SV 40* genome multiplies several times in S phase reaching a high replication titer. Therefore, *SV40* escapes the control mechanisms restricting *X. laevis* and human replication to once per cell cycle.

4.2 X. laevis as a Model System – Advantages and Limitations

By hormonal manipulation female frogs can be stimulated to lay eggs. The eggs are big (1.1–1.3 mm) in comparison to human cells and an additionally high egg number gives a good source for extract preparation and protein purification. Furthermore, most *X. laevis* proteins are very similar in sequence and behavior to their human counterparts, allowing easy knowledge transfer.

X. laevis eggs are naturally arrested in M phase of meiosis II. This allows starting experiments with a synchronized extract. *X. laevis* extracts can be used with a wide range of biochemical methods investigating replication. It is for example, possible to immunodeplete proteins, to see if this protein is required for replication and/or to investigate its binding partners. It is also common to add drugs to the extract to investigate some changes of proteins in their chromatin binding.

Furthermore, a *X. laevis* S phase takes roughly 60 min and is much shorter than a typical human S phase of several hours. *X. laevis* extract is also capable of replicating all kinds of DNA templates and replication is limited to once per cell cycle.

However, the sequence independent initiation of replication might be a problem for some experiments. Also, the lack of gap phases in the early embryo might suggest that there are less regulatory mechanisms in place. *X. laevis* is a model system not only in the field of DNA replication but also to study early embryo development. However, it is not suitable for genetics due to the tetraploid genome of *X. laevis*.

5 SV40 and Disease

SV40 is a polyomavirus. The virus is potentially oncogenic [multiple (poly-) tumors (-oma)], but so far there are not strong enough data to conclude that *SV40* has a causal role in human neoplasia. On the other hand, in some human tumors the presence of *SV40* DNA and gene expression has been proven. A potential transformation of a healthy cell into a cancerous one results as a consequence of the interaction of the three regulatory proteins: large T-antigen, small t-antigen and agnoprotein with cellular signaling proteins, regulating the cell cycle, DNA repair, and other processes (White et al. 2005). The potential of *SV40* DNA to be integrated into cellular DNA opened up a new research area, developing vectors derived from wild-type *SV40* in order to fight disease.

Acknowledgments. We thank the Health research Board (HRB) for funding this project.

References

Bailis, J. M. and S. L. Forsburg (2004). "MCM proteins: DNA damage, mutagenesis and repair." Curr Opin Genet Dev **14**(1): 17–21.

Bergsma, D. J., D. M. Olive, S. W. Hartzell and K. N. Subramanian (1982). "Territorial limits and functional anatomy of the simian virus 40 replication origin." Proc Natl Acad Sci U S A **79**(2): 381–5.

Blow, J. J. and A. Dutta (2005). "Preventing re-replication of chromosomal DNA." Nat Rev Mol Cell Biol **6**(6): 476–86.

Blow, J. J., P. J. Gillespie, D. Francis and D. A. Jackson (2001). "Replication origins in Xenopus egg extract are 5–15 kilobases apart and are activated in clusters that fire at different times." J Cell Biol **152**(1): 15–25.

Borowiec, J. A. and J. Hurwitz (1988). "Localized melting and structural changes in the SV40 origin of replication induced by T-antigen." Embo J **7**(10): 3149–58.

Burkhart, R., D. Schulte, D. Hu, C. Musahl, F. Gohring and R. Knippers (1995). "Interactions of human nuclear proteins P1Mcm3 and P1Cdc46." Eur J Biochem **228**(2): 431–8.

Bylund, G. O., J. Majka and P. M. Burgers (2006). "Overproduction and purification of RFC-related clamp loaders and PCNA-related clamps from *Saccharomyces cerevisiae*." Methods Enzymol **409**: 1–11.

Chong, J. P., H. M. Mahbubani, C. Y. Khoo and J. J. Blow (1995). "Purification of an MCM-containing complex as a component of the DNA replication licensing system." Nature **375**(6530): 418–21.

Collins, K. L. and T. J. Kelly (1991). "Effects of T antigen and replication protein A on the initiation of DNA synthesis by DNA polymerase alpha-primase." Mol Cell Biol **11**(4): 2108–15.

Costanzo, V., D. Shechter, P. J. Lupardus, K. A. Cimprich, M. Gottesman and J. Gautier (2003). "An ATR- and Cdc7-dependent DNA damage checkpoint that inhibits initiation of DNA replication." Mol Cell **11**(1): 203–13.

Cvetic, C. and J. C. Walter (2005). "Eukaryotic origins of DNA replication: could you please be more specific?" Semin Cell Dev Biol **16**(3): 343–53.

Danis, E., K. Brodolin, S. Menut, D. Maiorano, C. Girard-Reydet and M. Mechali (2004). "Specification of a DNA replication origin by a transcription complex." Nat Cell Biol **6**(8): 721–30.

Dean, F. B. and J. Hurwitz (1991). "Simian virus 40 large T antigen untwists DNA at the origin of DNA replication." J Biol Chem **266**(8): 5062–71.

Devault, A., E. A. Vallen, T. Yuan, S. Green, A. Bensimon and E. Schwob (2002). "Identification of Tah11/Sid2 as the ortholog of the replication licensing factor Cdt1 in Saccharomyces cerevisiae." Curr Biol **12**(8): 689–94.

Diffley, J. F. (2004). "Regulation of early events in chromosome replication." Curr Biol **14**(18): R778–86.

Donovan, S., J. Harwood, L. S. Drury and J. F. Diffley (1997). "Cdc6p-dependent loading of Mcm proteins onto pre-replicative chromatin in budding yeast." Proc Natl Acad Sci U S A **94**(11): 5611–6.

Dornreiter, I., W. C. Copeland and T. S. Wang (1993). "Initiation of simian virus 40 DNA replication requires the interaction of a specific domain of human DNA polymerase alpha with large T antigen." Mol Cell Biol **13**(2): 809–20.

Edwards, M. C., A. V. Tutter, C. Cvetic, C. H. Gilbert, T. A. Prokhorova and J. C. Walter (2002). "MCM2–7 complexes bind chromatin in a distributed pattern surrounding the origin recognition complex in Xenopus egg extracts." J Biol Chem **277**(36): 33049–57.

Forsburg, S. L. (2004). "Eukaryotic MCM proteins: beyond replication initiation." Microbiol Mol Biol Rev **68**(1): 109–31.

Gai, D., R. Roy, C. Wu and D. T. Simmons (2000). "Topoisomerase I associates specifically with simian virus 40 large-T-antigen double hexamer-origin complexes." J Virol **74**(11): 5224–32.

Garg, P. and P. M. Burgers (2005). "DNA polymerases that propagate the eukaryotic DNA replication fork." Crit Rev Biochem Mol Biol **40**(2): 115–28.

Gilbert, D. M. (2004). "In search of the holy replicator." Nat Rev Mol Cell Biol **5**(10): 848–55.

Gilbert, D. M., H. Miyazawa and M. L. DePamphilis (1995). "Site-specific initiation of DNA replication in Xenopus egg extract requires nuclear structure." Mol Cell Biol **15**(6): 2942–54.

Gillespie, P. J., A. Li and J. J. Blow (2001). "Reconstitution of licensed replication origins on Xenopus sperm nuclei using purified proteins." BMC Biochem **2**: 15.

Harland, R. M. and R. A. Laskey (1980). "Regulated replication of DNA microinjected into eggs of Xenopus laevis." Cell **21**(3): 761–71.

Hartwell, L. H. (1973). "Three additional genes required for deoxyribonucleic acid synthesis in Saccharomyces cerevisiae." J Bacteriol **115**(3): 966–74.

Harvey, K. J. and J. Newport (2003). "CpG methylation of DNA restricts prereplication complex assembly in Xenopus egg extracts." Mol Cell Biol **23**(19): 6769–79.

Hsiao, C. L. and J. Carbon (1979). "High-frequency transformation of yeast by plasmids containing the cloned yeast ARG4 gene." Proc Natl Acad Sci U S A **76**(8): 3829–33.

Huberman, J. A. and A. D. Riggs (1968). "On the mechanism of DNA replication in mammalian chromosomes." J Mol Biol **32**(2): 327–41.

Hubscher, U., H. P. Nasheuer and J. E. Syvaoja (2000). "Eukaryotic DNA polymerases, a growing family." Trends Biochem Sci **25**(3): 143–7.

Hyrien, O., K. Marheineke and A. Goldar (2003). "Paradoxes of eukaryotic DNA replication: MCM proteins and the random completion problem." Bioessays **25**(2): 116–25.

Hyrien, O., C. Maric and M. Mechali (1995). "Transition in specification of embryonic metazoan DNA replication origins." Science **270**(5238): 994–7.

Iizuka, M., T. Matsui, H. Takisawa and M. M. Smith (2006). "Regulation of replication licensing by acetyltransferase Hbo1." Mol Cell Biol **26**(3): 1098–108.

Ishimi, Y. (1997). "A DNA helicase activity is associated with an MCM4, -6, and -7 protein complex." J Biol Chem **272**(39): 24508–13.

Iyer, L. M., D. D. Leipe, E. V. Koonin and L. Aravind (2004). "Evolutionary history and higher order classification of AAA+ ATPases." J Struct Biol **146**(1–2): 11–31.

Jacob, F. and S. Brenner (1963). "[On the regulation of DNA synthesis in bacteria: the hypothesis of the replicon.]." C R Hebd Seances Acad Sci **256**: 298–300.

Jares, P., A. Donaldson and J. J. Blow (2000). "The Cdc7/Dbf4 protein kinase: target of the S phase checkpoint?" EMBO Rep **1**(4): 319–22.

Jares, P., M. G. Luciani and J. J. Blow (2004). "A Xenopus Dbf4 homolog is required for Cdc7 chromatin binding and DNA replication." BMC Mol Biol **5**: 5.

Joo, W. S., H. Y. Kim, J. D. Purviance, K. R. Sreekumar and P. A. Bullock (1998). "Assembly of T-antigen double hexamers on the simian virus 40 core origin requires only a subset of the available binding sites." Mol Cell Biol **18**(5): 2677–87.

Kao, H. I. and R. A. Bambara (2003). "The protein components and mechanism of eukaryotic Okazaki fragment maturation." Crit Rev Biochem Mol Biol **38**(5): 433–52.

Kaplan, D. L., M. J. Davey and M. O'Donnell (2003). "Mcm4,6,7 uses a "pump in ring" mechanism to unwind DNA by steric exclusion and actively translocate along a duplex." J Biol Chem **278**(49): 49171–82.

Kim, H. Y., B. A. Barbaro, W. S. Joo, A. E. Prack, K. R. Sreekumar and P. A. Bullock (1999). "Sequence requirements for the assembly of simian virus 40 T antigen and the T-antigen origin binding domain on the viral core origin of replication." J Virol **73**(9): 7543–55.

Koonin, E. V. (1993). "A common set of conserved motifs in a vast variety of putative nucleic acid-dependent ATPases including MCM proteins involved in the initiation of eukaryotic DNA replication." Nucleic Acids Res **21**(11): 2541–7.

Koonin, E. V. (1993). "A superfamily of ATPases with diverse functions containing either classical or deviant ATP-binding motif." J Mol Biol **229**(4): 1165–74.

Korner, U., M. Bustin, U. Scheer and R. Hock (2003). "Developmental role of HMGN proteins in Xenopus laevis." Mech Dev **120**(10): 1177–92.

Kubota, Y., S. Mimura, S. Nishimoto, T. Masuda, H. Nojima and H. Takisawa (1997). "Licensing of DNA replication by a multi-protein complex of MCM/P1 proteins in Xenopus eggs." Embo J **16**(11): 3320–31.

Kubota, Y., Y. Takase, Y. Komori, Y. Hashimoto, T. Arata, Y. Kamimura, H. Araki and H. Takisawa (2003). "A novel ring-like complex of Xenopus proteins essential for the initiation of DNA replication." Genes Dev **17**(9): 1141–52.

Labib, K., J. A. Tercero and J. F. Diffley (2000). "Uninterrupted MCM2-7 function required for DNA replication fork progression." Science **288**(5471): 1643–7.

Laskey, R. A. and M. A. Madine (2003). "A rotary pumping model for helicase function of MCM proteins at a distance from replication forks." EMBO Rep **4**(1): 26–30.

Lee, J. K. and J. Hurwitz (2000). "Isolation and characterization of various complexes of the minichromosome maintenance proteins of Schizosaccharomyces pombe." J Biol Chem **275**(25): 18871–8.

Lei, M., Y. Kawasaki and B. K. Tye (1996). "Physical interactions among Mcm proteins and effects of Mcm dosage on DNA replication in Saccharomyces cerevisiae." Mol Cell Biol **16**(9): 5081–90.

Leonhardt, H., H. P. Rahn and M. C. Cardoso (1998). "Intranuclear targeting of DNA replication factors." J Cell Biochem Suppl **30-31**: 243–9.

Lundgren, M., A. Andersson, L. Chen, P. Nilsson and R. Bernander (2004). "Three replication origins in Sulfolobus species: synchronous initiation of chromosome replication and asynchronous termination." Proc Natl Acad Sci U S A **101**(18): 7046–51.

Mahbubani, H. M., J. P. Chong, S. Chevalier, P. Thommes and J. J. Blow (1997). "Cell cycle regulation of the replication licensing system: involvement of a Cdk-dependent inhibitor." J Cell Biol **136**(1): 125–35.

Mahbubani, H. M., T. Paull, J. K. Elder and J. J. Blow (1992). "DNA replication initiates at multiple sites on plasmid DNA in Xenopus egg extracts." Nucleic Acids Res **20**(7): 1457–62.

Maiorano, D., J. Moreau and M. Mechali (2000). "XCDT1 is required for the assembly of pre-replicative complexes in Xenopus laevis." Nature **404**(6778): 622–5.

Maric, C., M. Benard and G. Pierron (2003). "Developmentally regulated usage of Physarum DNA replication origins." EMBO Rep **4**(5): 474–8.

Matsumoto, T., T. Eki and J. Hurwitz (1990). "Studies on the initiation and elongation reactions in the simian virus 40 DNA replication system." Proc Natl Acad Sci U S A **87**(24): 9712–6.

McGarry, T. J. and M. W. Kirschner (1998). "Geminin, an inhibitor of DNA replication, is degraded during mitosis." Cell **93**(6): 1043–53.

Mills, A. D., J. J. Blow, J. G. White, W. B. Amos, D. Wilcock and R. A. Laskey (1989). "Replication occurs at discrete foci spaced throughout nuclei replicating in vitro." J Cell Sci **94 (Pt 3)**: 471–7.

Montagnoli, A., R. Bosotti, F. Villa, M. Rialland, D. Brotherton, C. Mercurio, J. Berthelsen and C. Santocanale (2002). "Drf1, a novel regulatory subunit for human Cdc7 kinase." Embo J **21**(12): 3171–81.

Murakami, Y., C. R. Wobbe, L. Weissbach, F. B. Dean and J. Hurwitz (1986). "Role of DNA polymerase alpha and DNA primase in simian virus 40 DNA replication in vitro." Proc Natl Acad Sci U S A **83**(9): 2869–73.

Nakamura, H., C. Wu, J. Kuang, C. Larabell and L. D. Etkin (2000). "XCS-1, a maternally expressed gene product involved in regulating mitosis in Xenopus." J Cell Sci **113 (Pt 13)**: 2497–505.

Neuwald, A. F., L. Aravind, J. L. Spouge and E. V. Koonin (1999). "AAA+: A class of chaperone-like ATPases associated with the assembly, operation, and disassembly of protein complexes." Genome Res **9**(1): 27–43.

Nguyen, V. Q., C. Co and J. J. Li (2001). "Cyclin-dependent kinases prevent DNA re-replication through multiple mechanisms." Nature **411**(6841): 1068–73.

Oehlmann, M., A. J. Score and J. J. Blow (2004). "The role of Cdc6 in ensuring complete genome licensing and S phase checkpoint activation." J Cell Biol **165**(2): 181–90.

Oku, T., S. Ikeda, H. Sasaki, K. Fukuda, H. Morioka, E. Ohtsuka, H. Yoshikawa and T. Tsurimoto (1998). "Functional sites of human PCNA which interact with p21 (Cip1/Waf1), DNA polymerase delta and replication factor C." Genes Cells **3**(6): 357–69.

Parsons, R., M. E. Anderson and P. Tegtmeyer (1990). "Three domains in the simian virus 40 core origin orchestrate the binding, melting, and DNA helicase activities of T antigen." J Virol **64**(2): 509–18.

Pavlov, Y. I., C. Frahm, S. A. Nick McElhinny, A. Niimi, M. Suzuki and T. A. Kunkel (2006). "Evidence that errors made by DNA polymerase alpha are corrected by DNA polymerase delta." Curr Biol **16**(2): 202–7.

Pestryakov, P. E., K. Weisshart, B. Schlott, S. N. Khodyreva, E. Kremmer, F. Grosse, O. I. Lavrik and H. P. Nasheuer (2003). "Human replication protein A. The C-terminal RPA70 and the central RPA32 domains are involved in the interactions with the 3′-end of a primer-template DNA." J Biol Chem **278**(19): 17515–24.

Prokhorova, T. A. and J. J. Blow (2000). "Sequential MCM/P1 subcomplex assembly is required to form a heterohexamer with replication licensing activity." J Biol Chem **275**(4): 2491–8.

Robinson, N. P., I. Dionne, M. Lundgren, V. L. Marsh, R. Bernander and S. D. Bell (2004). "Identification of two origins of replication in the single chromosome of the archaeon Sulfolobus solfataricus." Cell **116**(1): 25–38.

Rowles, A., J. P. Chong, L. Brown, M. Howell, G. I. Evan and J. J. Blow (1996). "Interaction between the origin recognition complex and the replication licensing system in Xenopus." Cell **87**(2): 287–96.

Silva, T., R. H. Bradley, Y. Gao and M. Coue (2006). "Xenopus CDC7/DRF1 complex is required for the initiation of DNA replication." J Biol Chem **281**(17): 11569–76.

Simmons, D. T., D. Gai, R. Parsons, A. Debes and R. Roy (2004). "Assembly of the replication initiation complex on SV40 origin DNA." Nucleic Acids Res **32**(3): 1103–12.

Snyder, M., W. He and J. J. Zhang (2005). "The DNA replication factor MCM5 is essential for Stat1-mediated transcriptional activation." Proc Natl Acad Sci U S A **102**(41): 14539–44.

Stinchcomb, D. T., K. Struhl and R. W. Davis (1979). "Isolation and characterisation of a yeast chromosomal replicator." Nature **282**(5734): 39–43.

Stinchcomb, D. T., M. Thomas, J. Kelly, E. Selker and R. W. Davis (1980). "Eukaryotic DNA segments capable of autonomous replication in yeast." Proc Natl Acad Sci U S A **77**(8): 4559–63.

Tada, S., A. Li, D. Maiorano, M. Mechali and J. J. Blow (2001). "Repression of origin assembly in metaphase depends on inhibition of RLF-B/Cdt1 by geminin." Nat Cell Biol **3**(2): 107–13.

Takahashi, T. S. and J. C. Walter (2005). "Cdc7–Drf1 is a developmentally regulated protein kinase required for the initiation of vertebrate DNA replication." Genes Dev **19**(19): 2295–300.

Tanaka, S. and J. F. Diffley (2002). "Interdependent nuclear accumulation of budding yeast Cdt1 and Mcm2-7 during G1 phase." Nat Cell Biol **4**(3): 198–207.

Trowbridge, P. W., R. Roy and D. T. Simmons (1999). "Human topoisomerase I promotes initiation of simian virus 40 DNA replication in vitro." Mol Cell Biol **19**(3): 1686–94.

Tuteja, N. and R. Tuteja (2004). "Prokaryotic and eukaryotic DNA helicases. Essential molecular motor proteins for cellular machinery." Eur J Biochem **271**(10): 1835–48.

Van Hatten, R. A., A. V. Tutter, A. H. Holway, A. M. Khederian, J. C. Walter and W. M. Michael (2002). "The Xenopus Xmus 101 protein is required for the recruitment of Cdc45 to origins of DNA replication." J Cell Biol **159**(4): 541–7.

Voitenleitner, C., C. Rehfuess, M. Hilmes, L. O'Rear, P. C. Liao, D. A. Gage, R. Ott, H. P. Nasheuer and E. Fanning (1999). "Cell cycle-dependent regulation of human DNA polymerase alpha-primase activity by phosphorylation." Mol Cell Biol **19**(1): 646–56.

Waga, S., G. Bauer and B. Stillman (1994). "Reconstitution of complete SV40 DNA replication with purified replication factors." J Biol Chem **269**(14): 10923–34.

Waga, S. and A. Zembutsu (2006). "Dynamics of DNA binding of replication initiation proteins during de novo formation of pre-replicative complexes in Xenopus egg extracts." J Biol Chem **281**(16): 10926–34.

Walter, J. and J. Newport (2000). "Initiation of eukaryotic DNA replication: origin unwinding and sequential chromatin association of Cdc45, RPA, and DNA polymerase alpha." Mol Cell **5**(4): 617–27.

Walter, J. C. (2000). "Evidence for sequential action of cdc7 and cdk2 protein kinases during initiation of DNA replication in Xenopus egg extracts." J Biol Chem **275**(50): 39773–8.

Warbrick, E. (2000). "The puzzle of PCNA's many partners." Bioessays **22**(11): 997–1006.

Weisshart, K., H. Forster, E. Kremmer, B. Schlott, F. Grosse and H. P. Nasheuer (2000). "Protein–protein interactions of the primase subunits p58 and p48 with simian virus 40 T antigen are required for efficient primer synthesis in a cell-free system." J Biol Chem **275**(23): 17328–37.

Weisshart, K., P. Pestryakov, R. W. Smith, H. Hartmann, E. Kremmer, O. Lavrik and H. P. Nasheuer (2004). "Coordinated regulation of replication protein A activities by its subunits p14 and p32." J Biol Chem **279**(34): 35368–76.

White, M. K., J. Gordon, K. Reiss, L. Del Valle, S. Croul, A. Giordano, A. Darbinyan and K. Khalili (2005). "Human polyomaviruses and brain tumors." Brain Res Brain Res Rev **50**(1): 69–85.

Woodward, A. M., T. Gohler, M. G. Luciani, M. Oehlmann, X. Ge, A. Gartner, D. A. Jackson and J. J. Blow (2006). "Excess Mcm2-7 license dormant origins of replication that can be used under conditions of replicative stress." J Cell Biol **173**(5): 673–83.

Yanow, S. K., D. A. Gold, H. Y. Yoo and W. G. Dunphy (2003). "Xenopus Drf1, a regulator of Cdc7, displays checkpoint-dependent accumulation on chromatin during an S-phase arrest." J Biol Chem **278**(42): 41083–92.

Yoshizawa-Sugata, N., A. Ishii, C. Taniyama, E. Matsui, K. Arai and H. Masai (2005). "A second human Dbf4/ASK-related protein, Drf1/ASKL1, is required for efficient progression of S and M phases." J Biol Chem **280**(13): 13062–70.

You, Z., Y. Ishimi, H. Masai and F. Hanaoka (2002). "Roles of Mcm7 and Mcm4 subunits in the DNA helicase activity of the mouse Mcm4/6/7 complex." J Biol Chem **277**(45): 42471–9.

Chapter 2
A Genetic Screen Implicates miRNA-372 and miRNA-373 as Oncogenes in Testicular Germ Cell Tumors

P. Mathijs Voorhoeve,[1,5] Carlos le Sage,[1,5] Mariette Schrier,[1] Ad J.M. Gillis,[2] Hans Stoop,[2] Remco Nagel,[1] Ying-Poi Liu,[1] Josyanne van Duijse,[1] Jarno Drost,[1] Alexander Griekspoor,[1] Eitan Zlotorynski,[1] Norikazu Yabuta,[4] Gabriella De Vita,[3] Hiroshi Nojima,[4] Leendert H.J. Looijenga,[2] and Reuven Agami[1,*]

[1]*Division of Tumour Biology, The Netherlands Cancer Institute, Amsterdam, The Netherlands, *Contact: r.agami@nki.nl*
[2]*Department of Pathology, Erasmus University Medical Center Rotterdam, Daniel den Hoed Cancer Center, Josephine Nefkens Institute, Rotterdam, The Netherlands*
[3]*Dipartimento di Biologia e Patologia Cellulare e Molecolare, Università Federico II, CEINGE Biotecnologie Avanzate, Naples, Italy*
[4]*Department of Molecular Genetics, Osaka University, Japan*
[5]*These authors contributed equally to this work*

Abstract. Endogenous small RNAs (miRNAs) regulate gene expression by mechanisms conserved across metazoans. While the number of verified human miRNAs is still expanding, only few have been functionally annotated. To perform genetic screens for novel functions of miRNAs, we developed a library of vectors expressing the majority of cloned human miRNAs and created corresponding DNA barcode arrays. In a screen for miRNAs that cooperate with oncogenes in cellular transformation, we identified miR-372 and miR-373, each permitting proliferation and tumorigenesis of primary human cells that harbor both oncogenic RAS and active wild-type p53. These miRNAs neutralize p53-mediated CDK inhibition, possibly through direct inhibition of the expression of the tumorsuppressor LATS2. We provide evidence that these miRNAs are potential novel oncogenes participating in the development of human testicular germ cell tumors by numbing the p53 pathway, thus allowing tumorigenic growth in the presence of wild-type p53.

1 Introduction

Since their discovery, the functions of only a handful of microRNAs (miRNAs) have been determined (recently reviewed in Zamore and Haley, 2005). Relevant to carcinogenesis, it was found that let-7 inhibits RAS expression and in lung tumors negatively correlates with RAS levels (Johnson et al., 2005). Furthermore, the oncogenic potential of the miR-17-92 cluster was demonstrated (He et al., 2005; O'Donnell et al., 2005). This cluster is amplified in lymphomas (Ota et al., 2004), and its introduction accelerates tumorigenicity by an as yet undefined process. These findings demonstrate the powerful ability of small RNAs to alter cellular pathways and programs. However, the small number of miRNAs with a known function stresses the need for a systematic screening approach to identify more miRNA functions.

The difficulties in deciphering the mechanism of action of miRNAs with a known function and deducing their activity from their sequence are largely due to the complex relationship with their target genes. In general, target genes containing sequences that are completely complementary to the miRNA will be degraded by an RNA-interference mechanism, whereas targets with partial complementary sequences at their 3'-UTR will be subjected to translation inhibition and to a lesser extent also to mRNA degradation (Doench and Sharp, 2004; Bagga et al., 2005; Lim et al., 2005; Pillai et al., 2005). In mammals, a near-perfect complementarity between miRNAs and protein coding genes almost never exists, making it difficult to directly pinpoint relevant downstream targets of a miRNA. Several algorithms were developed that predict miRNA targets, most notably TargetScanS, PicTar, and miRanda (John et al., 2004; Lewis et al., 2005; Robins et al., 2005). These programs predict dozens to hundreds of target genes per miRNA, making it difficult to directly infer the cellular pathways affected by a given miRNA. Furthermore, the biological effect of the downregulation depends greatly on the cellular context, which exemplifies the need to deduce miRNA functions by in vivo genetic screens in well-defined model systems.

The cancerous process can be modeled by in vitro neoplastic transformation assays in primary human cells (Hahn et al., 1999). Using this system, sets of genetic elements required for transformation were identified. For example, the joint expression of the telomerase reverse transcriptase subunit (hTERT), oncogenic H-RASV12, and SV40-small t-antigen combined with the suppression of p53 and p16INK4A were sufficient to render primary human fibroblasts tumorigenic (Voorhoeve and Agami, 2003). Recently, these neoplastic transformation assays were used to uncover novel human tumor suppressor genes (Kolfschoten et al., 2005; Westbrook et al., 2005).

Moreover, oncogenes such as H-RASV12 provoke a stress response in primary cells that results in an irreversible growth arrest, termed premature senescence (Serrano et al., 1997). The senescent phenotype was recently shown to play a role in the protection from tumor development in vivo (Braig et al., 2005; Chen et al., 2005; Collado et al., 2005; Michaloglou et al., 2005).

The elimination of this protective mechanism by, for example, the suppression of the p53 and p16INK4A pathways permits continued proliferation of the modified primary cells in the presence of the oncogenic event, consequently leading to tumorigenicity (Voorhoeve and Agami, 2003). Here, we use this model system to perform a functional genetic screen to identify miRNAs that act as oncogenes in tumorigenesis. We characterize two miRNAs whose expression can substitute for the loss of wild-type (wt) p53 that is needed to overcome oncogene-mediated arrest and implicate their involvement in the formation of testicular germ cell tumors.

2 Results

miR-Vec: A vector-based miRNA expression system to identify novel functions of miRNAs, we constructed a retroviral vector for miRNA expression (miR-Vec) following a previously described approach (Chen et al., 2004). We inserted 1,500 bp fragments spanning a given miRNA-genomic region in a modified pMSCV-Blasticidin vector such that they are placed under the control of a CMV promoter (Figure 1A). To examine miRNA expression from the miR-Vec system, a miR-24 minigene-containing virus was transduced into human cells. Expression was determined using an RNase protection assay (RPA) with a probe designed to identify both precursor and mature miR-24 (Figure 1B). Figure 1C shows that cells transduced with miR-Vec-24 clearly express high levels of mature miR-24, whereas little expression was detected in control transduced cells. Furthermore, we confirmed the consistency of miRNA expression driven by miR-Vec by cloning eight miR-Vec plasmids expressing randomly chosen miRNAs. With one exception, all constructs yielded high expression levels of mature miRNAs (Figure S1). Notably, very little pre-miRNA accumulation was detected in all cases, indicating the efficient processing of the ectopically expressed miRNAs in the cells.

Next, we examined the functionality of the miR-Vec system to suppress gene expression by using both GFP, tagged with a sequence complementary to miR-19, and luciferase containing either the wt 3′-UTR of G6PD, a predicted miR-1 target, or control with two mutated miR-1 binding sequences (Lewis et al., 2003). Using fluorescence microscopy and luciferase assays, we observed potent and specific miRNA activity expressed from each miR-Vec (Figure S2). These results demonstrate the general applicability of miR-Vec to drive functional miRNA expression. We subsequently created a human miRNA expression library (miR-Lib) by cloning almost all annotated human miRNAs into our vector (Rfam release 6) (Figure S3). Additionally, we made a corresponding microarray (miR-Array) containing all miR-Lib inserts, which allows the detection of miRNA effects on proliferation.

To test the sensitivity of screens with miR-Lib and miRArray, we transduced modified primary BJ fibroblasts expressing ecotropic receptor and immortalized

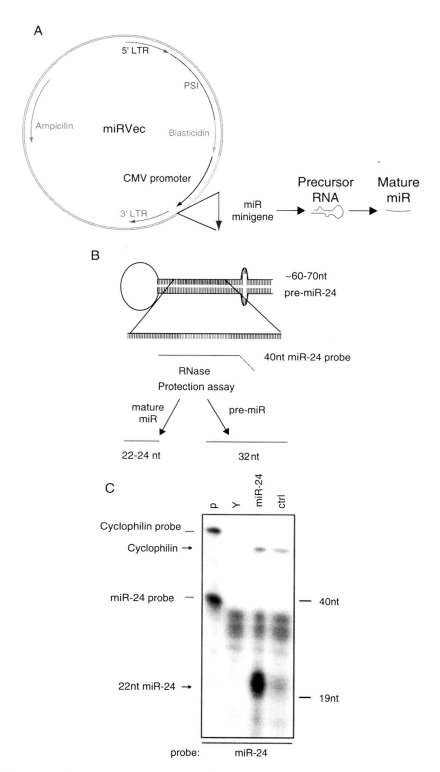

FIGURE 1. (*Continued*)

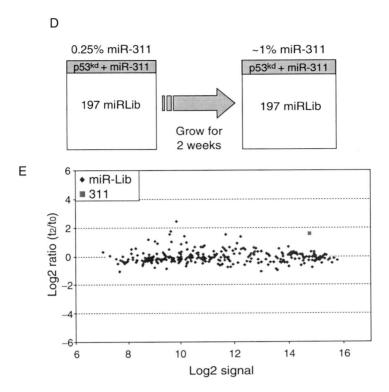

FIGURE 1. cont'd Tools for functional genetic screens with human miRNAs. (A) The miR-Vec miRNA-expressing system. Transcription of the minigene mimics the pri-miRNA, which is subsequently processed to a mature miRNA. (B) The RPA technique used to detect precursor and mature miRNAs in this study. (C) RPA was performed on RNA extracts from primary human BJ cells stably transduced with miR-Vec-24 and miR-Vec-ctrl. We used a probe to cyclophilin to control for loading. (P = 10% input probe, Y = Yeast control RNA). (D) and (E) BJ/ET-p53kd cells were transduced with miR-Vec-311, and BJ/ET cells were transduced with a mix of 197 other miR-Vecs. Both populations were drug selected, mixed in a ratio of 1:400, and left to grow for 2 weeks. A barcode experiment was done comparing cells right after mixing (t0) and after 2 weeks in culture (t2). The log2 of the ratio of the signals between t2 and t0 was plotted against the average signal to visualize outliers. The signal derived from the spot corresponding to Mir-311 is indicated in pink.

with hTERT (BJ/ET) with a mixture of 197 different miR-Vecs and mixed them in a ratio of 400:1 with BJ/ET cells containing both miR-Vec-311 and a knockdown construct for p53 (p53kd) (Figure 1D). Previously, we have shown that in a period of 2 weeks BJ/ET-p53kd cells increase four- to five fold in number compared with BJ/ET cells (Voorhoeve and Agami, 2003). In accordance, we observed an approximately fourfold increase in miR-311 signal, indicating that our procedure is sensitive enough to detect mild growth differences (Figure 1E).

2.1 Expression of miR-372 and miR-373 Protects from Oncogenic Stress

In response to mitogenic signals from oncogenes such as RASV12, primary human cells undergo a growth arrest (Figure 2A) (Serrano et al., 1997). In contrast, primary cells lacking functional p53 efficiently overcome this arrest. This escape from oncogene-induced senescence is a prerequisite for full transformation into tumor cells. To identify miRNAs that can interfere with this process and thus might contribute to the development of tumor cells, we transduced BJ/ET fibroblasts with miR-Lib and subsequently transduced them with either RASV12 or a control vector (Figure 2B). After 2 or 3 weeks in culture, senescence-induced differences in abundance of all miR-Vecs were determined with the miR-Array. Figures 2C and S3 show that in three independent experiments the relative abundance of three miR-Vecs increased reproducibly in the RASV12- expressing population. These hits corresponded to three constructs derived from one genomic region expressing miRNAs 371, 372, 373, and 373* (Figure 2D). Due to the proximity of miR-371 and miR-372 in the genome (within 0.5 kb), two largely overlapping constructs encoded both miRNA-371 and 372 (miR-Vec-371&2).

The third construct did not overlap with miR-Vec-371&2 and encoded miRNA-373 and 373*. Interestingly, the mature miR-373 is a homolog of miR-372, and neither shares obvious homology with either miRNA-371 or miR-373* (Figure 2D). This suggests that miR-372 and miR-373 caused the observed selective growth advantage. Next, we verified miRNA function in a cell-growth assay. First, we verified the expression of both miR-371 and 372 by miR-Vec-371&2 and miR-373 by miR-Vec-373 (Figure 2E). We then transduced BJ/ET cells with miR-Vec-371&2, miR-Vec-373, or the controls p53kd and vector control and then with RASV12. As expected, control cells ceased proliferating in response to RASV12, whereas p53kd cells continued to proliferate (Figure 2F). The expression of either miR-371&2 or miR-373 allowed cells to continue proliferating in the presence of oncogenic stress, validating the effect observed with the miR-Array. Oncogene-induced senescence is characterized by the appearance of cells with a flat morphology that express senescence associated (SA-β-Galactosidase). Indeed, control RASV12-arrested cells showed relatively high abundance of flat cells expressing SA-β-Galactosidase (Figure 2G, H). Consistent with the cell growth assay, very few cells showed senescent morphology when transduced with either miR-Vec-371&2, miR-Vec-373, or control p53kd. Altogether, these data show that transduction with either miR–Vec-371&2 or mi-R-Vec-373 prevents RASV12-induced growth arrest in primary human cells. The independent identification of constructs encoding two very similar miRNAs (miR-372 and miR-373) suggests that they (but not miR-371 or miR-373*) are required to cause this phenotype. To test this, we mutated the sequences

2. A Genetic Screen Implicates miRNA-372 and miRNA-373

A

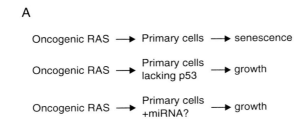

B

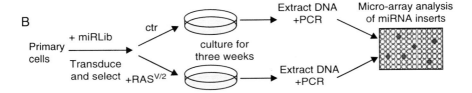

C

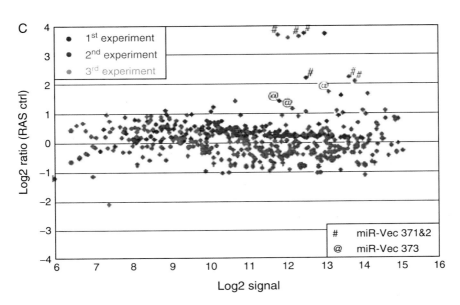

FIGURE 2. Identification of miR-Vecs that inhibit oncogene-induced senescence. (**A**) The effects of oncogenic RASV12 on cellular growth. (**B**) A flow chart of the screen. Cells transduced with the miR-Lib were grown for 2 to 3 weeks in the presence or absence of RASV12. Subsequently, the population of inserts in each condition was recovered and compared using miR-array. (**C**) Three independent miR-Array experiments were performed. The position of the reproducibly upregulated miR-Vecs is indicated for each experiment.

(*Continued*)

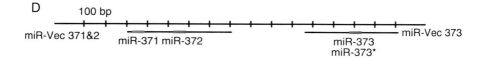

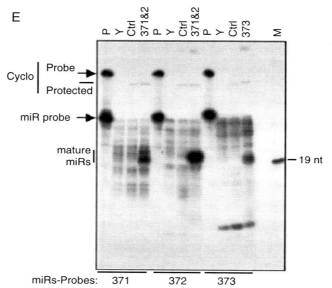
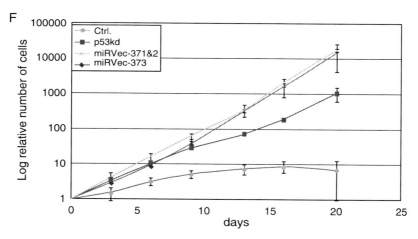

FIGURE 2. (*Continued*)

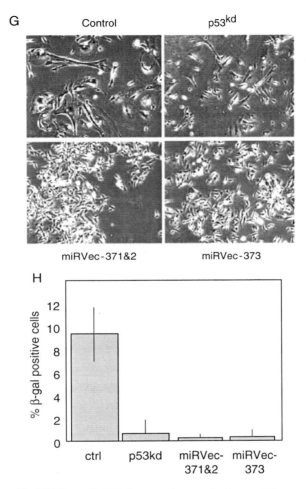

FIGURE 2. cont'd. (**D**) The miR-371–3 genomic organization and the sequences of the mature miRNAs expressed from this locus. For comparison, the nucleotides 2–8 (seed) of the miRNAs are boxed. (**E**) RPA analysis of RNA from BJ/ET cells containing the indicated miR-Vecs. (**F**) BJ/ET cells containing the indicated vectors were transduced with RASV12, drug selected, and subjected to a growth assay. Standard deviations from three independent transductions are shown. (**G**) The cells from (F) were stained 10 days after RASV12 transduction to detect SA-β-galactosidase expression. (**H**) The percentage of β-galactosidase positive cells was counted in three independent dishes.

of miR-372 and miR-373. As demonstrated by RPA, miR-Vec-372mut and miR-Vec-373mut indeed failed to express miR-372 and miR-373, respectively (Figure 3A).

Note that miR-Vec-372mut still expressed miR-371 to a similar extent as the original miR-Vec-371&2. We then tested these constructs in a YFP-competition assay to detect possible growth advantages conferred by the miRNAs on

BJ cells in the absence or presence of RASV12 (Figure 3B). For this purpose we used a miR-Vec vector that expresses YFP instead of a blasticidin resistance marker and compared the growth rates of YFP-tagged and untagged cells within one population. Increase in time of the YFP-positive cells within the population indicates a growth advantage conferred by the additional genetic unit encoded by the YFP vector. We also generated BJ/ET cells expressing the RASV12-ERTAM chimera gene, which is only active when tamoxifen is added (De Vita et al., 2005), and tranduced them with either YFP-tagged wt or mutant miR-Vec-371&2 and miR-Vec-373 constructs as well as p53kd, p14ARFkd, or control vectors. Figure 3B shows that even without activating RASV12 (no tamoxifen added), both miR-Vec-371&2 and miR-Vec-373 conferred a growth advantage to cells, although to a lesser extent than observed with p53kd or p14ARFkd. Once RASV12 was activated, the growth advantage of cells with miR-Vec-371&2 and 373 increased dramatically, indicating that these constructs allowed growth of cells in the presence of oncogenes while the rest of the population ceased to proliferate. In accordance with previously published data (Voorhoeve and Agami, 2003), reducing p53, but not p14ARF expression, was sufficient to overcome the oncogenic stress. Consistent with our assumption that miR-372&3 are the active miRNAs, mutating their mature sequence abrogated their growth advantage. This shows that miR-372&3, but not miR-371 or miR-373*, caused stimulation of proliferation and resistance to oncogenic stress.

2.2 Expression of miR-372 and miR-373 Transforms Primary Human Cells

Suppression of cellular senescence is essential for tumorigenesis. We therefore examined whether the ectopic expression of miR-372/3 is sufficient to replace loss of p53 in transformation of cells. A hallmark of cellular transformation is the ability of tumor cells to grow anchorage independently in semisolid medium and as tumors in model mice (Hahn et al., 1999; Hanahan and Weinberg, 2000). Indeed, in a soft agar assay, modified primary human BJ/ET cells expressing hTERT, SV40-small t, RASV12, and shRNA-knockdowns for p53 and p16INK4A showed potent ability to grow in an anchorage-independent manner (Figures 3C and 3D). To mimic the expression of the complete *miR-371–373* gene cluster, we made a miR-Vec expressing all miRNAs from one cluster (miR-Vec-cluster; see Figure 3A for expression). Similar to the knockdown of p53, the ability to grow in soft agar was also observed for cells containing miR-Vec-cluster, miR-Vec-371&2, or miR-Vec-373 but not miRVec-372mut or miR-Vec-373mut (Figure 3C, D). Moreover, the cells containing the miR-371–373 cluster grew efficiently as tumors in athymic nude mice (Figure 3E).

These results demonstrate that miR-372&3 collaborate with RAS in transformation in a manner that resembles p53 inactivation.

Importantly, our results so far indicate that the expression of miR-372&3 did not reduce the activity of RASV12, as these cells were still growing faster than normal cells and were tumorigenic, for which RAS activity is indispensable (Hahn et al., 1999; Kolfschoten et al., 2005). Therefore, the miRNA-mediated circumvention of the activation of p53 can in principle be obtained at a level upstream of p53, on p53 itself, or downstream. To shed more light on this aspect, we examined the effect of miR-372&3 expression on p53 activation in response to oncogenic stimulation. We used for this experiment

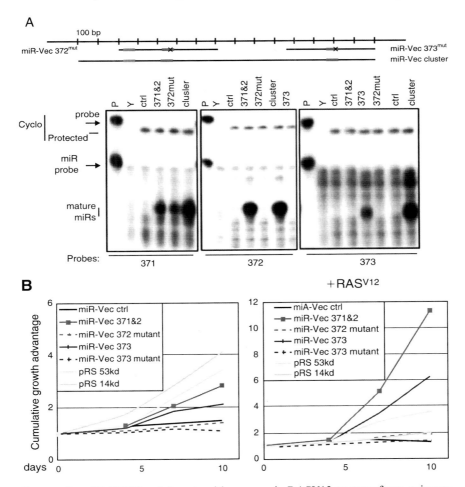

FIGURE 3. miR-372&3 collaborate with oncogenic RASV12 to transform primary human cells. (A) The mature miR-372&3 sequences were mutated in their corresponding miR-Vecs, and their expression was examined by RPA. miR-Vec cluster is a construct encompassing miRs-371–3. (B) The indicated YFP-containing vectors were transduced in BJ/ET-RASV12-ERTAM cells. The cumulative growth advantage was determined in the absence or presence of tamoxifen (+RASV12).

(*Continued*)

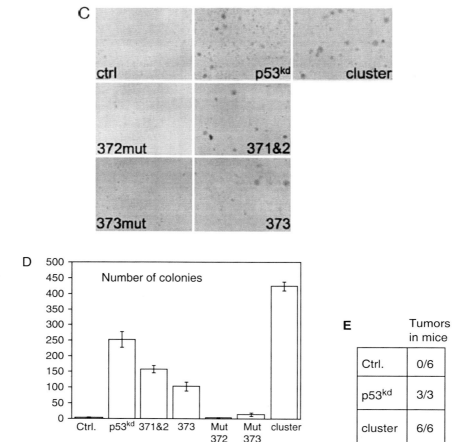

FIGURE 3. cont'd. (C) and (D) BJ/ET cells containing SV40 small t, p16INK4A knockdown, RASV12, and the indicated constructs were plated in soft agar, and colonies were photographed and counted after 3 weeks. The average and standard deviation of three independent dishes are shown. (E) The cluster, p53kd, and control cell populations from (C) were injected subcutaneously in athymic nude mice, and tumor growth was scored 5 weeks later.

BJ/ET cells containing p14ARFkd because, following RASV12 treatment, in those cells p53 is still activated but more clearly stabilized than in parental BJ/ET cells (Voorhoeve and Agami, 2003), resulting in a sensitized system for slight alterations in p53 in response to RASV12. Figure 4A shows that following RASV12 stimulation, p53 was stabilized and activated, and its target gene, *p21cip1*, was induced in all cases, indicating an intact p53 pathway in these cells. Therefore, it is unlikely that the miRNAs act on a factor upstream of p53 or on p53 itself to suppress the cellular response to oncogenic RAS.

Increased levels of p21cip1 inhibit CDK activity, causing cells to arrest in G1 phase (el-Deiry et al., 1993), whereas suppression of p21cip1 allows cells to grow in the presence of RASV12 (Figure S4A). To test whether p21cip1 was still functional in the miRNA-transduced cells, we examined CDK2 activity using an IP-Kinase assay (Figure 4B). In both miR-372&3-expressing cells, CDK2 remained active following RASV12 induction, whereas it was inhibited in the control cells (Figure 4B). In contrast, miR-372&3-transduced cells were still sensitive to inhibition of CDK activity by roscovitin (Figure S4B). This indicates that the presence of miR-372/3 acts as a molecular switch to make CDK2 resistant to increased levels of the cell cycle inhibitor p21cip1. Both p53 and p21cip1 play a major role in the DNA damage response to ionizing radiation (IR) (Weinert, 1998). Since cells expressing miR-372&3 proliferate in the presence of increased p21cip1 levels, we examined their response to damaged DNA. In the presence of both miRNAs, and irrespective of RASV12 expression, IR induced a cell cycle arrest that was indistinguishable from control cells, whereas the suppression of p53 expression allowed, as expected, continuous DNA replication (Figure 4C and data not shown). These results indicate that although miR-372&3 confer complete protection to oncogene-induced senescence in a manner similar to p53 inactivation, the cellular response to DNA damage remains intact.

2.3 *Potential Role of miR-372 and miR-373 in Human Cancer*

Based on the above results, we hypothesized that miRNA-372&3 may participate in tumorigenesis of some tumors that retain wt p53 and are sensitive to DNA-damaging treatments. One such tumor type is the testicular germ cell tumor of adolescents and adults (TGCT), known for the presence of wt p53 in the majority of cases and known to be generally sensitive to chemotherapies as well as irradiation (Kersemaekers et al., 2002; Masters and Koberle, 2003; Mayer et al., 2003). In addition, these tumors harbour an embryonic stem (ES) cell signature (Almstrup et al., 2004), which correlates with the reported ES-cell expression pattern of the miR-371–3 cluster (Suh et al., 2004). We therefore examined a number of cell lines originating from TGCTs for the expression of the miR-371–3 cluster. Four out of seven cell lines expressed this cluster (Figures 4D and S5). This result is significant as no clear expression of the miR-371–3 cluster was detected in any of the somatic cell lines we tested (originating from breast, colon, lung and brain tumors) (Figure S6).

TGCTs are divided into seminomas, nonseminomas, and spermatocytic-seminoma, according to their origin, clinical behavior, and chromosomal constitution (Oosterhuis and Looijenga, 2005). The nonseminomas can be composed of embryonal carcinoma (EC, the stem cell component), teratocarcinoma

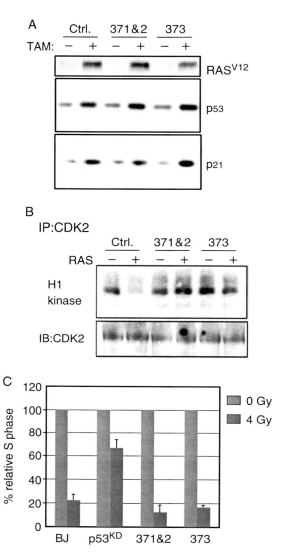

FIGURE 4. miR-372 and miR-373 sustain CDK2 activity in TGCT (**A**) BJ/ET-p14ARFkd-RASV12ERTAM cells containing the indicated miR-Vecs were cultured for a week in the presence or absence of tamoxifen (TAM), harvested, and subjected to immunoblot analyses to detect p21cip1, p53, and RASV12ER. (**B**) The same polyclonal populations as in (A) were harvested, and CDK2 kinase activity was measured using an IP-kinase protocol with Histone H1 as a substrate. Equal pulldown of CDK2 was checked by immunoblot of the same samples (lower panel). (**C**) BJ/ET cells containing either a control vector (BJ), or the indicated constructs were irradiated (4 Gy), labeled with BrdU, and subjected to flow cytometric analysis. The percentage of BrdU positive cells relative to the unirradiated cells is shown. SD is from three independent experiments.

2. A Genetic Screen Implicates miRNA-372 and miRNA-373

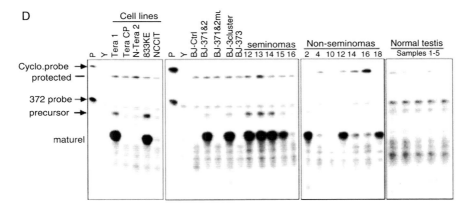

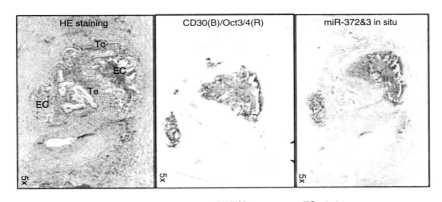

FIGURE 4. (Continued)

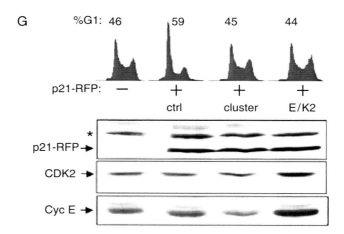

FIGURE 4. cont'd. (**D**) The expression of miR-372 was detected by RPA in RNA extracts from several TGCT cell lines as well as from primary seminoma and nonseminoma tumors and from normal testis tissues. (**E**) In situ hybridization on a nonseminoma of mixed histology to detect miR-372&3 expression. The probe was developed with NBT/BCIP blue, and the section was counterstained with FastRed. The EC component of the tumor was morphologically determined using HE-counter staining and by immunohistochemistry with anti-CD30 and OCT3/4 antibodies of the next sections. (**F**) Summary of the p53 status in several TGCT cell lines and primary seminomas (in the latter, only exons five to eight were examined). (**G**) NCCIT cells were cotransfected with H2B-GFP and the indicated constructs. Cell-cycle profiles of the GFP-positive population were examined after 4 days using flow cytometry. Also shown is an immunoblot analysis of cells from the same experiment with antibodies against p21cip1, CDK2, and cyclin E.

(TC, somatic differentiation), and yolk sac tumor, and choriocarcinoma (YS and CH, extra-embryonal tissues). All the cell lines we tested were derived from nonseminomatous tumors, as there are no other type of TGCT cell lines available so far. To substantiate our results and extend them to other TGCT types, we examined a panel of primary seminomas, nonseminomas, and spermatocytic seminomas for the expression of miR-372. Figures 4D and S7 show that most seminomas (28/32) had a clear miR-372 expression, that about two thirds (14/21) of the nonseminomas expressed miR-372, and that expression was observed in neither RNA from the spermatocytic-seminoma tumors (data not shown) nor from the normal testis tissue panel. Noteworthy is the fact that endogenous expression of miR-372 reached levels that are comparable to those driven by miR-Vec-372 (Figure 4D), indicating the biological relevance of our system in primary human fibroblast cells. Within the nonseminoma samples, both pure and mixed histologies were present (Figure S7). The RPA analysis showed that high expression of miR-372 correlated with a larger EC component. To further investigate this connection, we performed in situ miRNA hybridizations on tissue sections of 10 representative TGCTs and found in all cases 372&3 to be strictly localized to the EC component, as judged by morphology and immunohistochemistry with CD30 and Oct3/4 (Figure 4E and data not shown). Both seminomas and the EC component of nonseminomas share features with ES cells. To exclude that the detection of miR-371–3 merely reflects its expression pattern in ES cells, we tested by RPA miR-302a–d, another ES cells specific miRNA cluster (Suh et al., 2004). In many of the miR-371–3 expressing seminomas and nonseminomas, miR-302a–d was undetectable (Figures S7 and S8), suggesting that miR-371–3 expression is a selective event during tumorigenesis.

Interestingly, we noted a correlation between cluster expression and p53 status in the TGCT cell lines (Figure 4F). Whereas all three cluster-expressing cell lines contained high wt-p53 levels, NTera2 has low wild-type p53 levels, and NCCIT lost one p53 allele while the second allele is mutated (Burger et al., 1998). To strengthen the p53 connection seen in the TGCT cell lines, we examined p53 mutations in exons five to eight in the primary tumors, where the majority of mutations are found. In the nonseminoma panel, no mutations were detected. In contrast, two out of four miRNA 372-negative seminomas had an inactivating mutation in the *p53* gene (SE20 in exon 8 and SE28 in exon 5; Figures 4F and S7), a rare phenomenon in TGCT (Kersemaekers et al., 2002). In contrast, none of the 13 miRNA 372&373-expressing seminomas that we examined contained mutations in p53. Altogether, these results strongly suggest that the expression of miR-372/3 suppresses the p53 pathway to an extent sufficient to allow oncogenic mutations to accumulate in TGCTs. We then decided to test directly the correlation between the p53 pathway and miR-372/3 expression in TGCTs. It was technically not possible to sufficiently and persistently inhibit the expression of both miR-372 and miR-373 by methylated miRNA-oligos or knockdown vectors against the loop of the precursors (as judged by miR-372 and miR-373 luciferase reporter targets; data not shown). As an alternative approach, we used NCCIT, an embryonal carcinoma-derived cell line containing only a mutated, nonfunctional p53 that expresses very low amounts of the miR-371–3 cluster (Figure 4D) (Burger et al., 1998). We activated the p53 pathway by transfecting NCCIT cells with a p21-RFP construct that inhibits Cyclin E/CDK2 activity and examined the effects of miR372/3 by cotransfecting miR-Vec-371–3 (cluster). As expected, overexpression of p21-RFP caused accumulation of cells in G1, whereas cotransfection of Cyclin E/CDK2 allowed cells to continue proliferating in the presence of p21-RFP (Figure 4G). Significantly, cells cotransfected with the miR-Vec-cluster showed a phenotype similar to that observed with Cyclin E/CDK2. This result demonstrates the ability of miR-372 and miR-373 to overcome a p21-mediated cell cycle arrest in TGCTs and substantiates the correlation between these miRNAs, CDK and the p53 pathway.

2.4 miR-372 and miRNA-373 Regulate LATS2 Expression

Our results thus far indicate that miR-371–3 cluster suppresses an inhibitor of CDK activity and that this function is important for the development of TGCTs. To start to identify relevant targets of miR-372&3, we took advantage of the fact that miRNAs may cause limited destruction of their target mRNAs apart from inhibiting their translation (Lim et al., 2005). We performed an mRNA-expression array analysis comparing RASV12-expressing BJ/ET cells either containing p53kd or expressing the miR-371–3 cluster (Figures 5A and S9). We chose this setup as both cell types proliferate in the presence of oncogenic stress, thus canceling out the profound effects of cells going into senescence. We first looked in the p53kd cells and found p53 itself and many of its transcriptional targets to be downregulated compared to the cluster-expressing cells (Figure 5A). This independently confirms our previous results (Figure 4A) indicating that miR-372&373 do not directly inhibit p53 activity. From the list of genes whose expression was 2- or more fold lower in the cluster-expressing cells, we used target prediction programs to find possible direct targets of miR-372&3. We identified three miR-372/3 predicted targets: FYCO1 (FYVE and coiled coil containing protein 1), Suv39-H1, and LATS2 (Figure 5A). Interestingly, while nothing is known about the function of the FYCO1 protein, both Suv39-H1 and LATS2 have been connected in the past to RASV12-mediated transformation. It was recently shown that lymphocytes from mice nullizygous for Suv39-H1 are resistant to oncogene- induced senescence (Braig et al., 2005). However, the mechanism underlying this effect and its conservation to other tissues and to man are not known. Most promising seemed the LArge Tumor Suppressor homolog 2 (LATS2), a serine-threonine kinase whose deletion in flies accelerates cellular proliferation and tumorigenic development (Justice et al., 1995; Xu et al., 1995). In mice, a similar activity was seen in LATS2(–/–) mouse embryonic fibroblasts (McPherson et al., 2004), whereas its overexpression was shown to inhibit cyclin E/CDK2 activity and RASV12-mediated transformation (Li et al., 2003). Additionally, loss of LATS2 stimulated reduplication, an activity comparable to that observed when Cyclin E is overexpressed in the absence of p53 (Fukasawa et al., 1996; Tarapore and Fukasawa, 2002; Toji et al., 2004). Finally, in human breast cancer, hypermethylation of the LATS2 promoter was associated with an aggressive phenotype of the tumors (Takahashi et al., 2005). These observations suggest that the suppression of LATS2 explains at least in part the sustained activity of CDK in the presence of high p21cip1 levels in miR-372/3-expressing cells.

To investigate the possibility that miR-372 and miR-373 suppress the expression of LATS2, we performed immunoblot analysis of cells expressing wt and mutant miR- 372&3, the cluster and the controls p53kd and empty vector. Both in the absence of RASV12 and in its presence, a significant reduction in LATS2 protein level was observed upon miR-372&3 expression

(Figure 5B). Using quantitative RT-PCR and immunoblot analysis, we observed a two-fold effect on LATS2 RNA levels and four- to five-fold on protein levels by the miR-371–3 cluster (Figure 5C). As a control, we used an LATS2 knockdown construct (Figure 5F). These results show that a combined effect of RNA destruction and translation inhibition is used by miR-372&3 to silence LATS2.

miR-372/3 was predicted to bind two sites in the 3′-UTR of LATS2 that are highly conserved between human, mouse, and zebrafish (Figure 5D). To further substantiate LATS2 as a direct target of miR372&3, we cloned its 3′-UTR downstream of the firefly luciferase gene (*pGL3-LATS2*) (Figure 5E). We transfected either *pGL3-LATS2* or the controls *pGL3-372* and *pGL3-373* (containing a miR-complementary sequence in their 3′-UTR) or *pGL3* into Tera1 and MCF-7 cells (respectively positive and negative for miR-371–3) (Figures 4D and S6). As predicted, the 372/373 complementary sequences mediated strong inhibition of luciferase expression in Tera1 cells. Significantly, a potent inhibition of luciferase activity was also mediated by the 3′-UTR of LATS2 in either MCF-7 ectopically expressing miR-372 or in Tera1 cells but not by a construct mutated at both miR-372-predicted target sites. These results indicate that LATS2 is indeed a direct target of miR-372&3. Next, we tested whether LATS2 is a functional target of miR-372&3 using a YFP-competition assay. Indeed, inhibition of LATS2 conferred a growth advantage to cells expressing RASV12 (Figure 5F). The overall effect was less than the effect of the miR-Vec 373 but comparable to loss of p53. Therefore, these results point to LATS2 as a mediator of the miR-372 and miR-373 effects on cell proliferation and tumorigenicity, although they do not exclude the participation of other direct miR-targets, such as Suv39-H1, in these processes. Further investigation should elucidate the exact role of LATS2 downregulation and the possible participation of other miR-372&3-targets in the overall observed miR effect on cellular transformation.

3 Discussion

3.1 Functional Genetic Screens for miRNAs

We developed a miRNA-expression vector library and a corresponding barcode array to detect miRNAs whose expression modifies a defined cellular pathway. We demonstrate here the power of this technology by the identification of miRNA-372 and miRNA-373 as potential oncogenes that collaborate with oncogenic RAS in cellular transformation (Figure 5G). However, this strategy is also suitable for the identification of miRNAs that regulate other cellular pathways resulting in a proliferation or survival difference, such as the DNA damage response, differentiation, sensitivity to growth factors, and resistance to anticancer drugs. Furthermore, the miR-Lib tool can be used in a single-well format to identify growth-independent phenotypes.

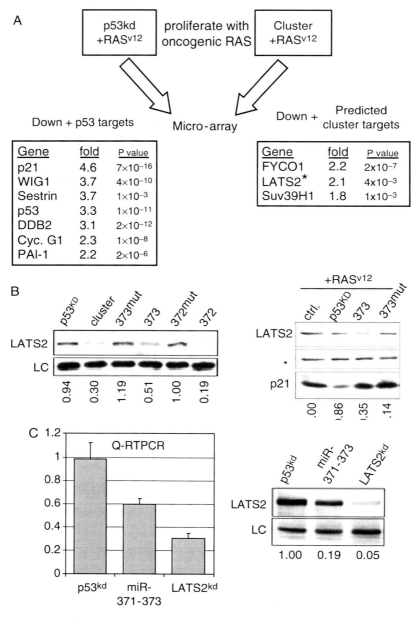

FIGURE 5. Inhibition of LATS2 expression by miR-372 and miR-373. (**A**) RNA was extracted from BJ/ET cells expressing RASV12 and containing either a p53kd construct or miR-Vec-371–3 (cluster) and compared using oligo-expression arrays (Figure S9). Listed are genes whose expression was downregulated in the p53kd cells and are known transcriptional targets of p53 as well as genes whose expression was suppressed in the cluster-expressing cells and are predicted TargetScanS targets of miRNA-372&3. LATS2*: The reduction was verified by Q-RT-PCR (Figure S10). (**B**) and (**C**) BJ/ET cells containing the indicated constructs were analyzed by immunoblot analysis or by Q-RT-PCR. Band intensity was calculated by densitometry.

(*Continued*)

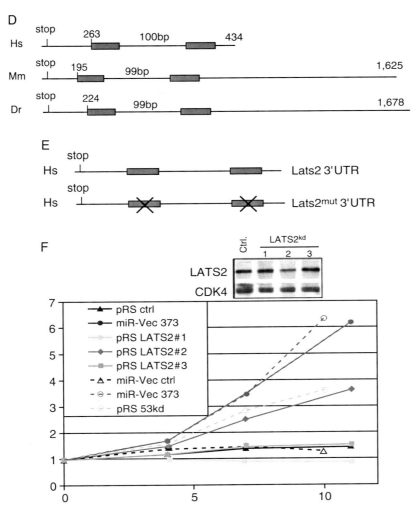

FIGURE 5. cont'd. (**D**) The 3′-UTRs of LATS2 in human (Hs), mouse (Mm), and zebrafish (Dr) are shown, and the predicted miR-372&3 target sequences are marked.(**E**) The indicated vectors were transfected in miR-372&3-positive (Tera1) and negative (MCF-7) cell lines. The relative firefly luciferase levels (divided to Renilla control and compared to pGL3) are shown. SD are from three independent experiments. (**F**) Cumulative growth advantage assay was performed as described in Figure 3B in RASV12-expressing cells transduced with the indicated vectors. For comparison, data from Figure 3B are included (*dashed lines*). An accompanying immunoblot shows that only LATS2kd#2 is functional.

(*Continued*)

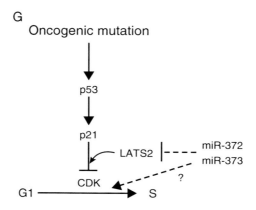

FIGURE 5. cont'd. (**G**) A schematic model showing the mechanism through which miR-372&3 can suppress an oncogene-activated *p53* pathway.

3.2 Collaboration of miRNA372&3 and RASV12 in Tumorigenesis

Sustained proliferation of cells in the presence of oncogenic signals is a major leap toward tumorigenicity (Hanahan and Weinberg, 2000). We found miR-372&3 to collaborate with RASV12 and stimulate a full-blown neoplastic transformation phenotype. However, whereas in the majority of the cases neoplastic transformation will require inactivation of p53 (for example by expression of HPV E6, HDM2, or mutant p53), miR372&3 uniquely allowed transformation to occur while p53 was active. This indicates that miRNA-372&3 do not block RASV12 signals but rather allow cells to proliferate irrespective of p53 activation and induction of p21cip1.

The expression of miR-372&3 results in prevention of the CDK inhibition that is caused by the oncogenic stress response. In both primary human fibroblasts and in a TGCT-derived cell line, cells expressing miR-372&3 were insensitive to elevated levels of the cell-cycle inhibitor p21cip1. Although the exact mechanism responsible for this effect is still unclear, we suggest that suppression of LATS2 is an important factor. Indeed, the expression of LATS2 is directly controlled by miR-372&3, and its activity is important for RASV12-induced senescence. However, further investigation is required to demonstrate the exact mechanism of LATS2 action and whether there are other targets of miR-372&3, such as Suv39H1, that are relevant to this phenotype.

3.3 Correlations to Other miRNAs

Based on the seed sequence, the *miR-372&3* gene family also includes *miR-93* and *miR-302a–e*. As these may share a broad range of target genes, they may also share many functions. Indeed, preliminary results show that similar to *miR-372&3*, albeit with minor differences, both *miR-93* and *miR-302a–e* can effectively target the LATS2 3′-UTR and bypass oncogene-induced senescence (manuscript in preparation).

3.4 Role of miR372&3 in TGCT Development

Our results suggest that during transformation, the activities of miR-372&3 circumvented the need to mutate p53, leading to a DNA-damage-sensitive transformed phenotype. These characteristics of miR-372&3-transformed primary human cells therefore suggest a role in wt p53-tumor genotypes that are also sensitive to chemotherapies, including irradiation. Indeed, TGCTs conform to this profile (Masters and Koberle, 2003). This could for instance be a result of high mdm2 levels, as was previously suggested for mouse teratocarcinomas (Lutzker and Levine, 1996). However, by several criteria mouse teratocarcinomas are counterparts of human germ cell tumors of neonates and infants rather than TGCT (Oosterhuis and Looijenga, 2005). Indeed, while the first show high mdm2 expression levels, this was not demonstrated in TGCT (Mostert et al., 2000). Therefore, it is highly significant that we found that miR-372&3-expressing TGCTs did not contain mutated p53 alleles, whereas a subset of miR-371–3 negative primary TGCTs and cell lines did. Altogether, these provide a strong indication that there is no selective advantage to mutate p53 during TGCT development when the miR-371–3 cluster is expressed.

The potent role of miR-372&3 in cellular transformation and potentially in TGCT development raises the possibility that they may play a similar role in somatic tumors. To this end, we determined the expression of the miR-371–3 cluster in several distinct somatic tumor cell lines and found little evidence for their expression (Figure S6). Consistent with our results, clear miR-371–3 cluster expression was observed in only 1 out of 70 leukemia tumors examined by others (Lu et al., 2005). It therefore seems that miR-372&3 expression is a rare event in somatic tumors. Whether such a role can be seen with the other members of the miR-372/3 family remains an open possibility.

Although both miR-372&3 and miR-302a–e clusters are expressed in ES cells (Suh et al., 2004), the miRNA-302 cluster is not expressed in many of these primary seminomas and nonseminomas (Figure S7). It is therefore most likely that the expression of miR-371–3 in primary TGCT is not merely a remnant of their ES cell phenotype but rather a selective event during TGCT tumorigenesis.

3.5 Function of miRNA372&3 in Embryonic Stem Cells

Our results suggest a link between the expression of miR-372&3 in embryonic stem cells and their function in cellular proliferation in these cells. miR-372&3 may facilitate rapid growth of stem cells by suppressing the expression of CDK inhibitors. Intriguingly, *Drosophila* germ cells and mouse embryonic stem (ES) cells require miRNAs to proliferate (Forstemann et al., 2005;

Hatfield et al., 2005). The proliferation defect in *Drosophila* mutants that lack miRNAs could be alleviated by loss of dacapo, the *Drosophila* p21cip1 homolog (Forstemann et al., 2005; Hatfield et al., 2005).

Our results indicate that due to enhanced tolerance to oncogenic mutations, deregulated expression of miR-372&3 predisposes cells for accumulation of carcinogenic events. Thus, the expression of these miRNAs must be carefully controlled during differentiation to prevent progression to cancer. Which factors control miR-371–3 expression during differentiation and whether their activity is causally related to development of TGCTs remain to be explored. Nevertheless, our experiments stress the importance of a strong downregulation of factors that maintain rapid cell proliferation, as in the absence of this downregulation safeguard mechanisms against oncogene emergence are functionally impaired.

4 Experimental Procedures

4.1 Constructs

pMSCV-Blast and pMSCV-YFP were made by replacing the puromycin resistance marker of pMSCV-puro (Clontech) with a PCR product encoding the blasticidin resistance gene from cDNA6/TR (invitrogen) or YFP from pEYFP-N1 (Clontech), respectively. pRetrosuper (pRS)- Blast was generated by replacing the 3'-LTR from pMSCV-Blast with the 3'-LTR from pRS-Hyg (Voorhoeve and Agami, 2003).

miR-Vec-Ctrl was made by deleting the MCS and the PGK-promoter from pMSCV-Blast, followed by insertion of the CMV promoter from pcDNA-3.1+ and a stuffer DNA derived from the first 211 nt of hTR downstream of the resistance marker. miR-Vec-YFP was cloned similar to miR-Vec-Ctrl, only starting from pMSCV-YFP. pBabe-puro-RasV12 and pBabe-puro, pMSCV-GFP-st, pRS-GFP, pBabe-H2BGFP, pCMV-Cyclin E, pCMV-CDK2, and pBabe-RasV12ERTAM were described before (Voorhoeve and Agami, 2003; De Vita et al., 2005). p53kd, p16kd, p14ARFkd, p21cip1kd shRNA constructs were described before (Voorhoeve and Agami, 2003; Duursma and Agami, 2005). pMSCV-Blast RASV12-ERTAM was made by subcloning RASV12-ERTAM into pMSCV-Blast. p21-RFP was produced by cloning p21 to the N terminus of dsRFP. The constructs encoding Luciferase-30-G6PD wt and mut were a kind gift of David Bartel (Lewis et al., 2003).

The miRNA minigenes were PCR amplified from genomic human DNA, cloned downstream of the CMV promoter in miR-Vec, and sequence verified. The primers used for the genomic PCR amplification of the individual miRNA minigenes, the miR-Vec-cluster and the miR-Vec mutants, are listed in Figure S3. LATS2 knockdown constructs were cloned to pRETROSUPER (pRS)-YFP (Brummelkamp et al., 2002). Targeting sequences are shown in Figure S3.

4.2 miR-Array

Genomic DNA was isolated from BJ/EHT cells with the DNeasy Tissue Kit (Qiagen). The inserts were recovered by PCR using primers listed in Figure S3. The PCR product was purified, and 500 ng was labeled using ULS-Cy3 or Cy5 (Kreatech) and hybridized to the miR-Array according to the manufacturers instructions (http://microarrays.nki.nl). As the amount of spots was too small to normalize automatically, the red and green signals were normalized by hand in Excel. For each spot, the log 2 of the red and green ratio as well as the log 2 of the square root of the product of the two signals were calculated. Outliers were picked, listed and compared across three independent experiments.

4.3 miRNA Detection

RNase Protection assays were performed using the mirVana miRNA probe construction and detection kits (Ambion) according to the manufacturer's instructions. 2.5–10 µg of RNA was used per reaction. Primers to make the RPA probes are listed in Figure S3. The antisense cyclophilin probe contained nucleotides 46–149 of Accession # BC013915. In situ hybridizations were performed with a mix of LNA oligos against miR-372 and miR-373 (Exiqon) according to the manufacturer's instructions.

4.4 Cell Culture and Antibodies

Primary BJ fibroblasts with an ecotropic receptor Neo and pBabepuro-hTert (BJ-ET) (Voorhoeve and Agami, 2003) or pBabe-H2BGFPhTert (BJ-EHT) (Kolfschoten et al., 2005) were grown in DMEM plus 10% FCS and antibiotics. NCCIT cells were grown in RPMI plus 10% FCS and antibiotics. Retrovirus was made by calcium-phosphate transduction of Eco-Pack 2 (Clontech) and harvesting 40 and 64 h later. BJ cells were selected with the relevant selective medium 48 h after transduction for at least a week. In the case of RASV12-encoding retroviruses, the selection was continued for the entire duration of the experiment. Antibodies used were DO-1 (p53), F5 (p21cip1), F235 (RAS), M20 (cyclin E), M2 (CDK2) from Santa Cruz Biotechnology, and 3D10 (LATS2) (Toji et al., 2004). Western blots were scanned and quantified using AIDA software (Raytek, Sheffield, UK).

4.5 Genetic Screen

BJ-EHT cells were transduced with a mixture of 197 miR-Vec vectors, drug selected for a week, and transduced independently three times with pBabe-puro-RASV12 or pBabe-puro. Cells from the independent transductions were propagated for 2 or 3 weeks before genomic DNA was isolated.

4.6 Growth Assay

BJ-EHT cells were transduced with miR-Vec or pRS-blast constructs, drug selected for a week, transduced with pBabe-Puro-RASV12, and drug selected for 3 days. 3×10^5 cells were plated in triplicates in 6 cm dishes and propagated twice a week. SA-β-galactosidase activity was assessed 10 days after RASV12 transduction, as described (Kolfschoten et al., 2005). 3×200 cells were scored for β-galactosides activity.

4.7 Soft Agar Assay and Tumorigenic Growth in Mice

BJ-EHT cells were transduced to more than 80% with pMSCV-GFP-st, pRS-Hyg-p16kd, drug selected, transduced with the various miR-Vec retroviruses or pRS-Blast-p53kd, drug selected again, and transduced with pBabe-puro-RASV12. After a week, the cells were either plated in triplicates in soft agar and macroscopically visible colonies were counted after 3 weeks or 10^6 cells were injected subcutaneously to athymic nude mice.

4.8 Cumulative Growth Advantage Assay

BJ/ET cells were transduced with pMSCV-Blast-RASV12-ERTAM, drugselected, and transduced with miR-Vec-YFP or pRS-GFP constructs. Efficiency of transduction (starting at 20–60%) was assessed by FACS in FL1, and cells were plated with and without $10'''7$ M 4-OHTTamoxifen. Cells were propagated, and the percentage of positive cells was measured twice a week. The relative growth advantage was calculated as described (Voorhoeve and Agami, 2003).

4.9 IP-Kinase Assay and Flow Cytometry

IP-kinase assay and flowcytometry were performed as described in Agami and Bernards (2000).

4.10 Expression Array Analysis and Target Prediction

Total RNA from BJ-EHT-st-p16kd-RASV12 cells either expressing a p53kd shRNA or the miR-Vec-cluster was extracted using Trizol (Invitrogen) and hybridized to an oligomicroarray using a standard protocol (http://microarrays.nki.nl). The genes that decreased 2-fold or more were further screened for possible miR-372/3 target sites using a local version of the TargetScan algrithm (Lewis et al., 2003) with default parameters (http://www.mekentosj.com/targetscanner).

Supplemental Data

Supplemental Data include seven figures and three tables and can be found with this article online at http://www.cell.com/cgi/content/full/ 124/6/1169/DC1/.

Acknowledgments. We thank Martijn Kedde and Hugo Horlings for technical help, Ron Kerkhoven and Mike Heimerikx for support in microarray analysis, Steve de Jong for reagents, Wigard Kloosterman for help in miRNAin situ protocol, J. Wolter Oosterhuis for supportive work for histology, and Alexandra Pietersen for critical reading of the manuscript. This work was supported by grants from the Dutch Cancer Society to P.M.V., C.S., and R.A. and by the EURYI award to R.A.

References

Agami, R., and Bernards, R. (2000). Distinct initiation and maintenance mechanisms cooperate to induce G1 cell cycle arrest in response to DNA damage. Cell 102, 55–66.

Almstrup, K., Hoei-Hansen, C.E., Wirkner, U., Blake, J., Schwager, C., Ansorge, W., Nielsen, J.E., Skakkebaek, N.E., Rajpert-De Meyts, E., and Leffers, H. (2004). Embryonic stem cell-like features of testicular carcinoma in situ revealed by genome-wide gene expression profiling. Cancer Res. 64, 4736–4743.

Bagga, S., Bracht, J., Hunter, S., Massirer, K., Holtz, J., Eachus, R., and Pasquinelli, A.E. (2005). Regulation by let-7 and lin-4 miRNAs results in target mRNA degradation. Cell 122, 553–563.

Braig, M., Lee, S., Loddenkemper, C., Rudolph, C., Peters, A.H., Schlegelberger, B., Stein, H., Dorken, B., Jenuwein, T., and Schmitt, C.A. (2005). Oncogene-induced senescence as an initial barrier in lymphoma development. Nature 436, 660–665.

Brummelkamp, T.R., Bernards, R., and Agami, R. (2002). A system for stable expression of short interfering RNAs in mammalian cells. Science 296, 550–553.

Burger, H., Nooter, K., Boersma, A.W., Kortland, C.J., and Stoter, G. (1998). Expression of p53, Bcl-2 and Bax in cisplatin-induced apoptosis in testicular germ cell tumour cell lines. Br. J. Cancer 77, 1562–1567.

Chen, C.Z., Li, L., Lodish, H.F., and Bartel, D.P. (2004). MicroRNAs modulate hematopoietic lineage differentiation. Science 303, 83–86.

Chen, Z., Trotman, L.C., Shaffer, D., Lin, H.K., Dotan, Z.A., Niki, M., Koutcher, J.A., Scher, H.I., Ludwig, T., Gerald, W., et al. (2005). Crucial role of p53-dependent cellular senescence in suppression of Ptendeficient tumorigenesis. Nature 436, 725–730.

Collado, M., Gil, J., Efeyan, A., Guerra, C., Schuhmacher, A.J., Barradas, M., Benguria, A., Zaballos, A., Flores, J.M., Barbacid, M., et al. (2005). Tumour biology: senescence in premalignant tumours. Nature 436, 642.

De Vita, G., Bauer, L., da Costa, V.M., De Felice, M., Baratta, M.G., De Menna, M., and Di Lauro, R. (2005). Dose-dependent inhibition of thyroid differentiation by RAS oncogenes. Mol. Endocrinol. 19, 76–89.

Doench, J.G., and Sharp, P.A. (2004). Specificity of microRNA target selection in translational repression. Genes Dev. 18, 504–511.

Duursma, A., and Agami, R. (2005). p53-Dependent regulation of Cdc6 protein stability controls cellular proliferation. Mol. Cell. Biol. 25, 6937–6947.

el-Deiry, W.S., Tokino, T., Velculescu, V.E., Levy, D.B., Parsons, R., Trent, J.M., Lin, D., Mercer, W.E., Kinzler, K.W., and Vogelstein, B. (1993). WAF1, a potential mediator of p53 tumor suppression. Cell 75, 817–825.

Forstemann, K., Tomari, Y., Du, T., Vagin, V.V., Denli, A.M., Bratu, D.P., Klattenhoff, C., Theurkauf, W.E., and Zamore, P.D. (2005). Normal microRNA maturation and germ-line stem cell maintenance requires Loquacious, a double-stranded RNA-binding domain protein. PLoS Biol. 3, e236.

Fukasawa, K., Choi, T., Kuriyama, R., Rulong, S., and Vande Woude, G.F. (1996). Abnormal centrosome amplification in the absence of p53. Science 271, 1744–1747.

Hahn, W.C., Counter, C.M., Lundberg, A.S., Beijersbergen, R.L., Brooks, M.W., and Weinberg, R.A. (1999). Creation of human tumour cells with defined genetic elements. Nature 400, 464–468.

Hanahan, D., and Weinberg, R.A. (2000). The hallmarks of cancer. Cell 100, 57–70.

Hatfield, S.D., Shcherbata, H.R., Fischer, K.A., Nakahara, K., Carthew, R.W., and Ruohola-Baker, H. (2005). Stem cell division is regulated by the microRNA pathway. Nature 435, 974–978.

He, L., Thomson, J.M., Hemann, M.T., Hernando-Monge, E., Mu, D., Goodson, S., Powers, S., Cordon-Cardo, C., Lowe, S.W., Hannon, G.J., and Hammond, S.M. (2005). A microRNA polycistron as a potential human oncogene. Nature 435, 828–833.

John, B., Enright, A.J., Aravin, A., Tuschl, T., Sander, C., and Marks, D.S. (2004). Human MicroRNA targets. PLoS Biol. 2, e363.

Johnson, S.M., Grosshans, H., Shingara, J., Byrom, M., Jarvis, R., Cheng, A., Labourier, E., Reinert, K.L., Brown, D., and Slack, F.J. (2005). RAS is regulated by the let-7 microRNA family. Cell 120, 635–647.

Justice, R.W., Zilian, O., Woods, D.F., Noll, M., and Bryant, P.J. (1995). The *Drosophila* tumor suppressor gene warts encodes a homolog of human myotonic dystrophy kinase and is required for the control of cell shape and proliferation. Genes Dev. 9, 534–546.

Kersemaekers, A.M., Mayer, F., Molier, M., van Weeren, P.C., Oosterhuis, J.W., Bokemeyer, C., and Looijenga, L.H. (2002). Role of P53 and MDM2 in treatment response of human germ cell tumors. J. Clin. Oncol. 20, 1551–1561. 1180 Cell 124, 1169–1181.

Kolfschoten, I.G., van Leeuwen, B., Berns, K., Mullenders, J., Beijersbergen, R.L., Bernards, R., Voorhoeve, P.M., and Agami, R. (2005). A genetic screen identifies PITX1 as a suppressor of RAS activity and tumorigenicity. Cell 121, 849–858.

Lewis, B.P., Shih, I.H., Jones-Rhoades, M.W., Bartel, D.P., and Burge, C.B. (2003). Prediction of mammalian microRNA targets. Cell 115, 787–798.

Lewis, B.P., Burge, C.B., and Bartel, D.P. (2005). Conserved seed pairing, often flanked by adenosines, indicates that thousands of human genes are microRNA targets. Cell 120, 15–20.

Li, Y., Pei, J., Xia, H., Ke, H., Wang, H., and Tao, W. (2003). Lats2, a putative tumor suppressor, inhibits G1/S transition. Oncogene 22, 4398–4405.

Lim, L.P., Lau, N.C., Garrett-Engele, P., Grimson, A., Schelter, J.M., Castle, J., Bartel, D.P., Linsley, P.S., and Johnson, J.M. (2005). Microarray analysis shows that some microRNAs downregulate large numbers of target mRNAs. Nature 433, 769–773.

Lu, J., Getz, G., Miska, E.A., Alvarez-Saavedra, E., Lamb, J., Peck, D., Sweet-Cordero, A., Ebert, B.L., Mak, R.H., Ferrando, A.A., et al. (2005). MicroRNA expression profiles classify human cancers. Nature 435, 834–838.

Lutzker, S.G., and Levine, A.J. (1996). A functionally inactive p53 protein in teratocarcinoma cells is activated by either DNA damage or cellular differentiation. Nat. Med. 2, 804–810.

Masters, J.R., and Koberle, B. (2003). Curing metastatic cancer: lessons from testicular germ-cell tumours. Nat. Rev. Cancer 3, 517–525.

Mayer, F., Stoop, H., Scheffer, G.L., Scheper, R., Oosterhuis, J.W., Looijenga, L.H., and Bokemeyer, C. (2003). Molecular determinants of treatment response in human germ cell tumors. Clin. Cancer Res. 9, 767–773.

McPherson, J.P., Tamblyn, L., Elia, A., Migon, E., Shehabeldin, A., Matysiak-Zablocki, E., Lemmers, B., Salmena, L., Hakem, A., Fish, J., et al. (2004). Lats2/Kpm is required for embryonic development, proliferation control and genomic integrity. EMBO J. 23, 3677–3688.

Michaloglou, C., Vredeveld, L.C., Soengas, M.S., Denoyelle, C., Kuilman, T., van der Horst, C.M., Majoor, D.M., Shay, J.W., Mooi, W.J., and Peeper, D.S. (2005). BRAFE600-associated senescence-like cell cycle arrest of human naevi. Nature 436, 720–724.

Mostert, M., Rosenberg, C., Stoop, H., Schuyer, M., Timmer, A., Oosterhuis, W., and Looijenga, L. (2000). Comparative genomic and in situ hybridization of germ cell tumors of the infantile testis. Lab. Invest. 80, 1055–1064.

O'Donnell, K.A., Wentzel, E.A., Zeller, K.I., Dang, C.V., and Mendell, J.T. (2005). c-Myc-regulated microRNAs modulate E2F1 expression. Nature 435, 839–843.

Oosterhuis, J.W., and Looijenga, L.H. (2005). Testicular germ-cell tumours in a broader perspective. Nat. Rev. Cancer 5, 210–222.

Ota, A., Tagawa, H., Karnan, S., Tsuzuki, S., Karpas, A., Kira, S., Yoshida, Y., and Seto, M. (2004). Identification and characterization of a novel gene, C13orf25, as a target for 13q31-q32 amplification in malignant lymphoma. Cancer Res. 64, 3087–3095.

Pillai, R.S., Bhattacharyya, S.N., Artus, C.G., Zoller, T., Cougot, N., Basyuk, E., Bertrand, E., and Filipowicz, W. (2005). Inhibition of translational initiation by Let-7 MicroRNA in human cells. Science 309, 1573–1576.

Robins, H., Li, Y., and Padgett, R.W. (2005). Incorporating structure to predict microRNA targets. Proc. Natl. Acad. Sci. USA 102, 4006–4009.

Serrano, M., Lin, A.W., McCurrach, M.E., Beach, D., and Lowe, S.W. (1997). Oncogenic ras provokes premature cell senescence associated with accumulation of p53 and p16INK4a. Cell 88, 593–602.

Suh, M.R., Lee, Y., Kim, J.Y., Kim, S.K., Moon, S.H., Lee, J.Y., Cha, K.Y., Chung, H.M., Yoon, H.S., Moon, S.Y., et al. (2004). Human embryonic stem cells express a unique set of microRNAs. Develop. Biol. 270, 488–498.

Takahashi, Y., Miyoshi, Y., Takahata, C., Irahara, N., Taguchi, T., Tamaki, Y., and Noguchi, S. (2005). Down-regulation of LATS1 and LATS2 mRNA expression by promoter hypermethylation and its association with biologically aggressive phenotype in human breast cancers. Clin. Cancer Res. 11, 1380–1385.

Tarapore, P., and Fukasawa, K. (2002). Loss of p53 and centrosome hyperamplification. Oncogene 21, 6234–6240.

Toji, S., Yabuta, N., Hosomi, T., Nishihara, S., Kobayashi, T., Suzuki, S., Tamai, K., and Nojima, H. (2004). The centrosomal protein Lats2 is a phosphorylation target of Aurora-A kinase. Genes Cells 9, 383–397.

Voorhoeve, P.M., and Agami, R. (2003). The tumor-suppressive functions of the human INK4A locus. Cancer Cell 4, 311–319.

Weinert, T. (1998). DNA damage and checkpoint pathways: molecular anatomy and interactions with repair. Cell 94, 555–558.

Westbrook, T.F., Martin, E.S., Schlabach, M.R., Leng, Y., Liang, A.C., Feng, B., Zhao, J.J., Roberts, T.M., Mandel, G., Hannon, G.J., et al. (2005). A genetic screen for candidate tumor suppressors identifies REST. Cell 121, 837–848.

Xu, T., Wang, W., Zhang, S., Stewart, R.A., and Yu, W. (1995). Identifying tumor suppressors in genetic mosaics: the *Drosophila* lats gene encodes a putative protein kinase. Development 121, 1053–1063.

Zamore, P.D., and Haley, B. (2005). Ribo-gnome: the big world of small RNAs. Science 309, 1519–1524.

Chapter 3
An Investigation into 53BP1 Complex Formation

Kevin C. Roche[1] and Noel F. Lowndes[1]*

[1] *National University of Ireland Galway, Department of Biochemistry, kevin.roche@nuigalway.ie,*
*noel.lowndes@nuigalway.ie

Abstract. Loss of the control over cellular proliferation can lead to cell death or result in the abnormal proliferation characteristic of the cancerous state. Among the controls used to achieve normal cellular proliferation is the DNA damage checkpoint pathway that monitors genome integrity (Hartwell and Kastan 1994). 53BP1 was identified as a protein that interacts with the DNA-binding core domain of the tumor suppressor p53. The p53-binding region of 53BP1 maps to the C-terminal BRCT domains which are homologous to those found in the breast cancer protein BRCA1 and in other proteins involved in the DNA damage response, notably budding yeast Rad9. In addition to its recently reported role in sensing double strand breaks, 53BP1 is believed to have roles, currently ill understood, in many aspects of DNA metabolism ranging from transcription and class switch recombination to 'mediating' the DNA damage checkpoint response (Chai et al. 1999; Huyen et al. 2004; Sengupta et al. 2004; Ward et al. 2004). Here, we investigate 53BP1 complex formation. We investigate 53BP1 oligomerization and show that this is not dependent on the presence of disulfide bridges.

1 Introduction

The genome of the eukaryotic cell is under constant attack, whether from UV light, reactive chemicals, genotoxic agents or the intrinsic biochemical instability of the DNA itself. Fortunately, these attacks to DNA are largely counterbalanced by promptly deployed, multifaceted surveillance and rescue mechanisms (Bakkenist and Kastan 2003). These surveillance and rescue pathways can generally be described as cell cycle checkpoints. Cell cycle checkpoints are signal transduction pathways activated after DNA damage to protect genomic integrity (Rouse and Jackson 2002). The cell needs mechanisms to detect different types of DNA lesions and it must also be capable of detecting very low levels of DNA damage. Indeed, experiments in yeast show that a single persistent double strand break (DSB) can be detected (Lee et al. 1998). Therefore, the rapidity and potency of the DNA damage response indicates that the signalling proteins involved must be very sensitive and have the capacity to amplify the initial stimulus (Rouse and Jackson 2002).

1.1 ATM/ATR Checkpoint Kinases

At the heart of the DNA damage response are ATM and ATR – two central players in this highly conserved and regulated process. Homologs of ATM and ATR are present in all eukaryotic cell types examined, including budding and fission yeast. They belong to a structurally unique family of protein serine/threonine kinases termed phosphoinositide 3-kinase related kinases (PIKKs) (Abraham 2004). They appear to sit at the top of the signalling networks induced by DNA damage and from there orchestrate the damage response, sometimes together and sometimes separately. However, before describing their respective roles in checkpoint signalling we briefly discuss their biochemistry and regulation.

1.2 Biochemistry of the ATM and ATR Kinases

A number of studies have identified the consensus phosphorylation motif for ATM and ATR as being Ser/Thr-Gln (Kim et al. 1999; O'Neill, et al. 2000). Interestingly, all ATM, ATR and DNA-PK PIKKs characterised to-date, mammalian or otherwise, function as S/T-Q directed kinases. However, the similarity in terms of substrate preference contrasts sharply with the differential activities of these proteins during DNA damage responses. ATM and ATR respond to DNA damage in fundamentally different ways. ATR responds to DNA damage by undergoing a dramatic shift in intranuclear localization from being diffuse in the nucleus to localising into distinct nuclear foci. Under identical experiment conditions, ATM did not enter into nuclear foci in cells damaged with IR or other DNA damaging agents (Tibbetts et al. 2000), however, recent reports have suggested that activated ATM can form nuclear foci in particular when immunoflouresence studies are performed detecting the active form of ATM where serine 1981 is phosphorylated (Suzuki et al. 2006). It is also becoming clearer how these proteins are becoming activated. Recent evidence suggests, that ATM, in the absence of DNA damage, is present in the nucleus as a homodimer. Upon DSB formation, the two ATM proteins phosphorylate each other at Ser1981 and this event results in disruption of oligomerization and hence activation of ATM kinase activity (Bakkenist and Kastan 2003). Autophosphorylation of ATM has been shown in vitro, requiring only addition of ATP. Other nucleotides and DNA are not required, however, under the same conditions ATP failed to activate both ATR and DNA-PK, once again emphasising the different possible responses these proteins have to DNA damage (Kozlov et al. 2003).

1.3 The ATM/ATR-Associated Signalling Machinery

ATM and ATR are large proteins of approximately 350 and 305 kDa. They both contain an evolutionary conserved phosphoinositide 3-kinase domain in their C-terminus and large, ill-defined amino-termini, the function of which remains largely unknown (Abraham 2001). It has been speculated that these amino termini play a role mediating an interaction with scaffold proteins in macromolecular signalling complexes. Gel filtration analyses indicate that both ATM and ATR are constitutive residents of very high molecular weight protein complexes (>2 MDa) in mammalian cells (Wright et al. 1998). The identification of the composition of these complexes is an area of increasing interest. However, for the purposes of this report it is important to indicate the importance of the PIKK proteins in the regulation of the DNA damage response. In budding yeast, Mec1 (ATR homolog), plays a central role in the regulation of scRad9 and in mammalian systems both ATM and/or ATR have been shown to be activated by central players in the DNA damage response e.g., 53BP1, BRCA1 and MDC1 (Rappold et al. 2001; Mochan et al. 2003; Ouchi 2006) all of which are outlined below.

1.4 Rad9 – The First Checkpoint Protein

The PIKK kinases lie at the heart of the DNA damage response pathway and the accessory proteins, which localise to DNA lesions and regulate kinase activation, are conserved from yeast to mammals. In *Saccharomyces cerevisiae*, the DNA damage checkpoint is in a large part under the control of Mec1, the ortholog of ATR, together with its interacting protein, Ddc2. The Mec1–Ddc2 complex has been proposed to detect processed DNA lesions via an interaction with RPA-coated single stranded DNA (Zou and Elledge 2003). Upon Mec1 activation, the tandem BRCT domain-containing protein Rad9 becomes highly phosphorylated on putative Mec1 consensus sites characterised by Ser/Thr-Gln motifs (Emili 1998; Vialard et al. 1998; Schwartz et al. 2002). Rad9 phosphorylation triggers the binding of Rad53, the Chk2 ortholog, to Rad9 in a manner that depends on both the N-terminal (FHA1) and C-terminal (FHA2) FHA domains of Rad53 (Durocher and Jackson 1999). Rad9 is essential for DNA damage signalling in the G1 and G2/M phases of the cell cycle. However, in vertebrates there is no clear Rad9 ortholog readily identifiable but rather a family of tandem BRCT domain-containing proteins that act in the checkpoint response. These Rad9-like proteins include MDC1, BRCA1, and 53BP1.

1.5 MDC1 – Mediator of DNA Damage Checkpoint Protein 1

MDC1 (also termed NFBD1) has been proposed to be a protein which amplifies ATM-dependent DNA damage signals (Lou , et al. 2006). MDC1 contains an N-terminal FHA domain and two C-terminal BRCT domains which share homology with the C-terminal BRCT repeats of scRad9. It has a central region of 14 consecutive repeats of 41 amino acids which bind DNA-PKcs/Ku and facilitate DNA-PK-dependent DNA repair processes. However, MDC1 also performs roles early in the DNA damage response. Upon DNA damage, MDC1 is phosphorylated in an ATM dependent manner and rapidly localises to sites of damage. It is required for the localization of other DNA damage response proteins including Nbs1, 53BP1 and BRCA1 to nuclear foci following damage. This localization appears to be dependent on γ-H2AX as MDC1 does not form foci in $H2AX^{-/-}$ cells. Furthermore, MDC1 binds directly to phosphorylated but not to an unphosphorylated H2AX peptide in vitro (Stucki et al. 2005). Significantly, MDC1 knockout mice show a similar phenotype to $H2AX^{-/-}$ mice (Lou et al. 2006).

At the cellular level, $MDC1^{-/-}$ cells demonstrate a greatly reduced level of H2AX phosphorylation after damage although the initial γ-H2AX signal appears normal. This would point to a role for MDC1 not in the initial phosphorylation event but rather in the accumulation or maintenance of the signal. Consistent with this is the fact that ATM autophosphorylation is not affected in MDC1 deficient cells but the accumulation of this phosphorylated active ATM to sites of damage is abolished (Lou et al. 2006). It is also interesting to note that the binding of ATM to γ-H2AX in vitro and in vivo is dependent on MDC1, again, pointing to a role for MDC1 as a bridging molecule for ATM and H2AX. Finally, co-immunoprecipitation studies have shown that MDC1 interacts with ATM and γ-H2AX through its N-terminal FHA domain and its C-terminal BRCT domain respectively. It would appear that MDC1 is a true mediator in the amplification of DNA damage signals.

1.6 BRCA1 – Breast Cancer Susceptibility Gene Product 1

In the search for a mammalian homolog of Rad9, three proteins were found to share significant homology with the BRCT motifs of scRad9 – BRCA1, MDC1 and 53BP1. BRCA1 was identified in 1994 as the product of a gene that acted as a tumor suppressor in breast and ovarian cancer. Since then hundreds of mutations have been found in affected families and 80% of individuals carrying BRCA1 defects succumb to these diseases.

In terms of the DNA damage response, BRCA1 contains two C-terminal BRCT repeats which have homology with the BRCT domains of Rad9 from *S. cerevisiae* suggesting a role for BRCA1 in checkpoint signalling and DNA repair. It has also been demonstrated to have binding affinities with a large number of proteins that function in checkpoint activation and transcription, including the ATR kinase, PP1α protein phosphatase (Liu et al. 2002), SWI/SNF chromatin remodelling complexes (Bochar et al. 2000), the *c-myc* oncogene (Li et al. 2002) and the BACH1 helicase (Peng, et al. 2006).

BRCA1 has also been reported to bind DNA through a central domain spanning residues 452—1,079 with preferential binding to branched DNA structures. Although, BRCA1 could influence DNA repair indirectly through its role in transcription, its affinity for branched DNA and the MRN complex in particular suggesting that BRCA1 has direct roles in DSB repair (Durant and Nickoloff 2005).

Furthermore, defects in BRCA1 confer sensitivity to a broad range of DNA damaging agents suggesting that BRCA1 is a central component of the DNA damage response mechanism (Westermark et al. 2003; Zhang et al. 2004). It has also been shown that BRCA1 is required for ATM- and ATR-dependent phosphorylation of p53, c-Jun, Nbs1 and Chk2 following exposure to IR or UV radiation, respectively (Foray et al. 2003). However, BRCA1 appears to be dispensable for DNA damage induced phosphorylation of components of the checkpoint sliding clamp (hRad9, hRad1 and hHus1) as well as relocalization of this complex to DNA lesions. Foray et al. go on to propose that BRCA1 facilitates the ability of ATM and ATR to phosphorylate downstream substrates that directly influence cell cycle checkpoint arrest and apoptosis, but that BRCA1 is dispensable for the phosphorylation of DNA associated ATM and ATR substrates. These studies suggest that BRCA1 could be acting as a specific scaffold for the two PIKKs to phosphorylate a specific subset of their substrates.

As outlined, BRCA1 has been reported to have a number of roles in the DNA damage response, and its two C-terminal BRCT domains do share sequence homology with scRad9, however, recent reports about another putative ortholog-53BP1 suggest that this protein could have more functional homology to scRad9 than either BRCA1 or MDC1.

1.7 53BP1 – p53 Binding Protein 1

53BP1 was identified in a yeast two-hybrid screen seeking to find p53 interacting proteins. Two p53 interacting proteins were identified (53BP1 and 53BP2), however, their function does not appear to be related. Early studies demonstrated that 53BP1 was capable of enhancing p53 transcriptional activity and a model was proposed whereby 53BP1 was acting in a transcriptional coactivator role. However, 53BP1 was unusual in that it interacted with the central DNA binding region of p53 and not its N-terminal transactivation domain. This model of 53BP1 function became even more unlikely as the three dimensional 53BP1-p53 structure demonstrated that it is impossible for p53 to bind 53BP1 and p53 responsive promoters at the same time (Derbyshire et al. 2002).

Until recently, a tandem repeat of the BRCT motif found at the carboxyl terminus of 53BP1 was the only region of this protein with appreciable homology to other proteins. In fact, the BRCT domain of 53BP1 is most similar to the BRCT domains of BRCA1, MDC1(NFBD1), Rad9 and its fission yeast homolog, Crb2, defining a subclass of BRCT repeat proteins (Du et al. 2004). It was on this basis of sequence homology with the BRCT repeats of budding yeast Rad9 that 53BP1 was proposed to have a role in the DNA damage response, thus adding 53BP1 to the list of DNA damage checkpoint proteins listed above. Furthermore, a glycine-arginine rich (GAR) repeat region in 53BP1 has been identified and the methytransferase PRMT1 has been shown to asymmetrically methylate 53BP1, however, the exact function of this domain and modification remains unclear (Adams et al. 2005).

However, another recent study has described an additional conserved domain in 53BP1 involved in specific interaction with methylated lysine 79 of histone H3 (H3-K79me) that becomes exposed in the vicinity of double strand breaks (DSB) (Huyen et al. 2004). This domain consists of two tandem tudor folds forming the walls of a deep pocket that specifically interacts with H3-K79me. The domain is located in the central region of 53BP1 encompassing amino acids 1,487—1,532. Tudor domains generally consist of ~50–70 amino acids and are found in many eukaryotic proteins that colocalise with ribonucleoproteins or single-stranded DNA-associated complexes in the nucleus, mitochondrial membrane, or kinetochores. Importantly, this domain is also present in budding yeast Rad9 and fission yeast Crb2. Furthermore, this domain in Rad9 and Crb2 has been demonstrated to function in a similar manner to human 53BP1 (Huyen et al. 2004; Sanders et al. 2004; Wysocki et al. 2005) suggesting that these proteins are orthologs of 53BP1.

Thus, these proteins have related BRCT and Tudor domains and they all function in the DNA damage checkpoint pathways of three phylogenetically distant organisms. Moreover, all three proteins appear to be very similarly regulated, being hyperphosphorylated after DNA damage, with this regulation being dependent on the ATM/ATR/Mec1/Rad3[sp] class of phosphoinositide (PI3) kinases. These structural, functional and regulatory similarities are consistent with the possibility that 53BP1 and Rad9 are orthologs. However, despite these similarities determination of the precise biochemical function of these proteins will be required to fully address whether these proteins are indeed functional equivalents.

Previously, our laboratory has purified and characterised Rad9 containing protein complexes (Gilbert et al. 2001). We have identified two forms of Rad9 in soluble extracts: a large (>850 kDa) complex containing hypophosphorylated Rad9 and a smaller (560 kDa) complex containing hyperphosphorylated Rad9. We found that both forms co-purify with the Ssa1/2 chaperones, which we have genetically determined to be novel checkpoint activities required for Rad9 function (Gilbert et al. 2003). Rad53 only co-purifies with the smaller hyperphosphorylated Rad9 complex. Our data are consistent with the model that the larger Rad9 complex is remodelled after DNA damage, probably requiring Mec1-dependent hyperphosphorylation and chaperone activity. Rad53 is then recruited to the hyperphosphorylated Rad9 to form the 560 kDa complex (Gilbert et al. 2001). Once bound to the 560 kDa complex full activation of Rad53 by *in trans* autophosphorylation is facilitated. Furthermore, Rad9 (and its fission yeast ortholog, Crb2) has been shown to oligomerise through its BRCT repeats.

In addition to this, Adams et al. have reported that 53BP1 exists in a homo-oligomerised form, however, the precise nature of this oligomerization remains unclear. Here, we have investigated 53BP1 complex formation by performing experiments to demonstrate whether 53BP1 oligomerization is due to the presence of intermolecular disulfide bridges.

2 Materials and Methods

2.1 Whole Cell Lysates:

U2 OS cells were harvested in trypsin, spun at 1,000 rpm for 5 min and washed in PBS. Cell pellet was resuspended in whole cell extract buffer (WCEB – 20 mM Hepes pH7.6, 400 mM NaCl, 1 mM EDTA, 25% glycerol v/v, 1 mM DTT, 0.1% NP-40, phosphatase and protease inhibitors) and passed through a 25G needle 10 times. The extract was allowed to stand for 30 min and then spun for 15 min at 14,000 rpm. The extract was transferred to a different tube and protein content of the extract was determined by Bradford assay.

2.2 Western Blot

Proteins separated by SDS–PAGE were transferred to a Polyvinylidine difluoride membrane (Millipore). The membrane was soaked in transfer buffer solution prior to transfer and transfer carried out at 90v for 2 h. Membrane was blocked for 30 min in TBS-Tween supplemented with 10% w/v powdered milk, then probed with primary antibody overnight at 4°C. The membrane was washed in TBS-Tween 0.05% for 3 × 15 min, then exposed to secondary antibody for 45 min, RT and washed for 3 × 15 min in TBS-Tween 0.05%. 53BP1 antibodies were visualised using Supersignal® West Pico Chemiluminescent Substrate solution (Pierce) and exposed to Kodak X-OMAT AR film.

2.3 Antibodies

53BP1 antibody (abcam cat no. 4564).

3 Results and Discussion

As mentioned, Adams et al. have previously shown 53BP1 to oligomerize in vivo. Furthermore, they investigated what component of 53BP1 was responsible for the observed oligomerization. In a detailed study, they demonstrated that 53BP1 oligomerization was independent of methylation by PRMT1, its Tudor, GAR and BRCT domains and possibly independent of phosphorylation status. A region of 423 amino acids between residues 1,052 and 1,475 was identified as the domain that was responsible for homo-oligomerization of 53BP1. After looking at this region, we identified two cystine residues, from DSDBASE prediction program, within the oligomerization region which could act as a putative disulfide bridge and therefore be responsible for holding 53BP1 complexes together.

Extract from U2-OS cells was made in the absence of the reducing agent β-mercaptoethanol. This extract was then denatured in SDS loading buffer in the presence and absence of β-mercaptoethanol at varying temperatures. Samples were loaded onto the gel and a 53BP1 western blot was performed. (Figure 1). A gel with the same samples was stained with Coomassie Blue® to show equal loading.

No observable difference was detected between the 53BP1 in the extracts that were loaded in the presence of β-mercaptoethanol and those that were not (Figure 1). In the presence of an intermolecular disulfide bridge, one would expect 53BP1 to have a much higher mobility than the denatured and reduced band. Therefore, we can conclude that 53BP1 does not oligomerise due to the existence of intermolecular disulfide bridges. One would also expect that in the presence of an intramolecular disulfide bridge a conformational change in the protein would result causing it to run differently in an SDS–PAGE compared with a totally denatured and reduced 53BP1 (Figure 1 lane 7). However, no difference was detected between any of the different denaturing conditions suggesting that disulfide bridges do not play an important role in the folding or complex formation of 53BP1.

To further characterise 53BP1 complexes and to gain an insight into its similarities with Rad9 in terms of complex formation; gel filtration analysis needs to be performed. This analysis will give an indication of complex size and also demonstrate if the complex size changes after DNA damage. A different method that determines complex size would also be useful, for example, a glycerol gradient. Ultimately, however, purification of 53BP1 containing complexes is required to understand the true composition and size of 53BP1 complexes before and after DNA damage.

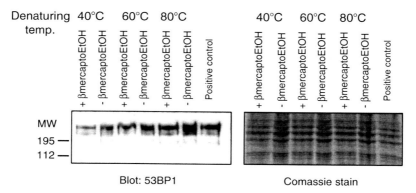

FIGURE 1. Denatured 53BP1 does not have an altered mobility in the absence of the strong reducing agent β-mercaptoethanol. 30 μg of protein were loaded in each lane. The positive lane indicates normal denaturing conditions of boiling in the presence of β-mercaptoethanol for 5 min.

4 Conclusion

53BP1 oligomerises in vivo (Adams et al. 2005). This study aimed to investigate whether this oligomerization was due to the presence of disulfide bridges between 53BP1 molecules. Our results suggest that intermolecular disulfide bridges are not responsible for 53BP1 oligomerization.

Acknowledgments. This work was supported by a fellowship to KCR from the Irish Research Council for Science, Engineering and Technology (IRCSET).

References

Abraham, R. T. (2001). "Cell cycle checkpoint signaling through the ATM and ATR kinases." Genes Develop **15**: 2177–96.

Abraham, R. T. (2004). "PI 3-kinase related kinases: 'big' players in stress-induced signaling pathways." DNA Repair (Amst) **3**(8–9): 883–7.

Adams, M. M., B. Wang, Z. Xia, J. C. Morales, X. Lu, L. A. Donehower, D. A. Bochar, S. J. Elledge and P. B. Carpenter (2005). "53BP1 oligomerization is independent of its methylation by PRMT1." Cell Cycle **4**(12): 1854–61.

Bakkenist, C. J. and M. B. Kastan (2003). "DNA damage activates ATM through intermolecular autophosphorylation and dimer dissociation." Nature **421**(6922): 499–506.

Bochar, D. A., L. Wang, H. Beniya, A. Kinev, Y. Xue, W. S. Lane, W. Wang, F. Kashanchi and R. Shiekhattar (2000). "BRCA1 is associated with a human SWI/SNF-related complex: linking chromatin remodeling to breast cancer." Cell **102**(2): 257–65.

Chai, Y. L., J. Cui, N. Shao, E. Shyam, P. Reddy and V. N. Rao (1999). "The second BRCT domain of BRCA1 proteins interacts with p53 and stimulates transcription from the p21WAF1/CIP1 promoter." Oncogene **18**(1): 263–8.

Derbyshire, D. J., B. P. Basu, L. C. Serpell, W. S. Joo, T. Date, K. Iwabuchi and A. J. Doherty (2002). "Crystal structure of human 53BP1 BRCT domains bound to p53 tumour suppressor." Embo J **21**(14): 3863–72.

Du, L. L., B. A. Moser and P. Russell (2004). "Homo-oligomerization is the essential function of the tandem BRCT domains in the checkpoint protein Crb2." J Biol Chem.

Durant, S. T. and J. A. Nickoloff (2005). "Good timing in the cell cycle for precise DNA repair by BRCA1." Cell Cycle **4**(9): 1216–22.

Durocher, D. H. J. F. A. R. and S. P. Jackson (1999). "The FHA domain is a modular phosphopeptide recognition motif." Mol Cell **4**(3): 387–94.

Emili, A. (1998). "MEC1-dependent phosphorylation of Rad9p in response to DNA damage." Mol Cell **2**(2): 183–9.

Foray, N., D. Marot, A. Gabriel, V. Randrianarison, A. M. Carr, M. Perricaudet, A. Ashworth and P. Jeggo (2003). "A subset of ATM- and ATR-dependent phosphorylation events requires the BRCA1 protein." Embo J **22**(11): 2860–71.

Gilbert, C. S., C. M. Green and N. F. Lowndes (2001). "Budding yeast Rad9 is an ATP-dependent Rad53 activating machine." Mol Cell **8**(1): 129–36.

Gilbert, C. S., M. van den Bosch, C. M. Green, J. E. Vialard, M. Grenon, H. Erdjument-Bromage, P. Tempst and N. F. Lowndes (2003). "The budding yeast Rad9 checkpoint complex: chaperone proteins are required for its function." EMBO Rep **4**(10): 953–8.

Hartwell, L. H. and M. B. Kastan (1994). "Cell cycle control and cancer." Science 266(5192): 1821–8.
Huyen, Y., O. Zgheib, R. A. Ditullio Jr, V. G. Gorgoulis, P. Zacharatos, T. J. Petty, E. A. Sheston, H. S. Mellert, E. S. Stavridi and T. D. Halazonetis (2004). "Methylated lysine 79 of histone H3 targets 53BP1 to DNA double-strand breaks." Nature 432(7015): 406–411.
Kim, S. T., D. S. Lim, C. E. Canman and M. B. Kastan (1999). "Substrate specificities and identification of putative substrates of ATM kinase family members." J Biol Chem 274(53): 37538–43.
Kozlov, S., N. Gueven, K. Keating, J. Ramsay and M. F. Lavin (2003). "ATP activates ATM in vitro: importance of autophosphorylation." J Biol Chem.
Lee, S. E., J. K. Moore, A. Holmes, K. Umezu, R. D. Kolodner and J. E. Haber (1998). "Saccharomyces Ku70, mre11/rad50 and RPA proteins regulate adaptation to G2/M arrest after DNA damage." Cell 94(3): 399–409.
Li, H., T. H. Lee and H. Avraham (2002). "A novel tricomplex of BRCA1, Nmi, and c-Myc inhibits c-Myc-induced human telomerase reverse transcriptase gene (hTERT) promoter activity in breast cancer." J Biol Chem 277(23): 20965–73.
Liu, Y., D. M. Virshup, R. L. White and L. C. Hsu (2002). "Regulation of BRCA1 phosphorylation by interaction with protein phosphatase 1alpha." Cancer Res 62(22): 6357–61.
Lou, Z., K. Minter-Dykhouse, S. Franco, M. Gostissa, M. A. Rivera, A. Celeste, J. P. Manis, J. van Deursen, A. Nussenzweig, T. T. Paull, F. W. Alt and J. Chen (2006). "MDC1 maintains genomic stability by participating in the amplification of ATM-dependent DNA damage signals." Mol Cell 21(2): 187–200.
Mochan, T. A., M. Venere, R. A. DiTullio, Jr. and T. D. Halazonetis (2003). "53BP1 and NFBD1/MDC1-Nbs1 function in parallel interacting pathways activating ataxia-telangiectasia mutated (ATM) in response to DNA damage." Cancer Res 63(24): 8586–91.
O'Neill, T., A. J. Dwyer, Y. Ziv, D. W. Chan, S. P. Lees-Miller, R. H. Abraham, J. H. Lai, D. Hill, Y. Shiloh, L. C. Cantley and G. A. Rathbun (2000). "Utilization of oriented peptide libraries to identify substrate motifs selected by ATM." J Biol Chem 275(30): 22719–27.
Ouchi, T. (2006). "BRCA1 Phosphorylation: Biological Consequences." Cancer Biol Ther 5(5): 470–5.
Peng, M., R. Litman, Z. Jin, G. Fong and S. B. Cantor (2006). "BACH1 is a DNA repair protein supporting BRCA1 damage response." Oncogene 25(15): 2245–53.
Rappold, I., K. Iwabuchi, T. Date and J. Chen (2001). "Tumor suppressor p53 binding protein 1 (53BP1) is involved in DNA damage-signaling pathways." J Cell Biol 153(3): 613–20.
Rouse, J. and S. P. Jackson (2002). "Interfaces between the detection, signaling, and repair of DNA damage." Science 297(5581): 547–51.
Sanders, S. L., M. Portoso, J. Mata, J. Bahler, R. C. Allshire and T. Kouzarides (2004). "Methylation of histone h4 lysine 20 controls recruitment of crb2 to sites of DNA damage." Cell 119(5): 603–14.
Schwartz, M. F., J. K. Duong, Z. Sun, J. S. Morrow, D. Pradhan and D. F. Stern (2002). "Rad9 phosphorylation sites couple Rad53 to the Saccharomyces cerevisiae DNA damage checkpoint." Mol Cell 9(5): 1055–65.
Sengupta, S., A. I. Robles, S. P. Linke, N. I. Sinogeeva, R. Zhang, R. Pedeux, I. M. Ward, A. Celeste, A. Nussenzweig, J. Chen, T. D. Halazonetis and C. C. Harris

(2004). "Functional interaction between BLM helicase and 53BP1 in a Chk1-mediated pathway during S-phase arrest." J Cell Biol **166**(6): 801–13.

Stucki, M., J. A. Clapperton, D. Mohammad, M. B. Yaffe, S. J. Smerdon and S. P. Jackson (2005). "MDC1 directly binds phosphorylated histone H2AX to regulate cellular responses to DNA double-strand breaks." Cell **123**(7): 1213–26.

Suzuki, K., H. Okada, M. Yamauchi, Y. Oka, S. Kodama and M. Watanabe (2006). "Qualitative and quantitative analysis of phosphorylated ATM foci induced by low-dose ionizing radiation." Radiat Res **165**(5): 499–504.

Tibbetts, R. S., D. Cortez, K. M. Brumbaugh, R. Scully, D. Livingston, S. J. Elledge and R. T. Abraham (2000). "Functional interactions between BRCA1 and the checkpoint kinase ATR during genotoxic stress." Genes Develop **14**(23): 2989–3002.

Vialard, J. E., C. S. Gilbert, C. M. Green and N. F. Lowndes (1998). "The budding yeast Rad9 checkpoint protein is subjected to Mec1/Tel1-dependent hyperphosphorylation and interacts with Rad53 after DNA damage." Embo J **17**(19): 5679–88.

Ward, I. M., B. Reina-San-Martin, A. Olaru, K. Minn, K. Tamada, J. S. Lau, M. Cascalho, L. Chen, A. Nussenzweig, F. Livak, M. C. Nussenzweig and J. Chen (2004). "53BP1 is required for class switch recombination." J Cell Biol **165**(4): 459–64.

Westermark, U. K., M. Reyngold, A. B. Olshen, R. Baer, M. Jasin and M. E. Moynahan (2003). "BARD1 participates with BRCA1 in homology-directed repair of chromosome breaks." Mol Cell Biol **23**(21): 7926–36.

Wright, J. A., K. S. Keegan, D. R. Herendeen, N. J. Bentley, A. M. Carr, M. F. Hoekstra and P. Concannon (1998). "Protein kinase mutants of human ATR increase sensitivity to UV and ionizing radiation and abrogate cell cycle checkpoint control." Proc Natl Acad Sci U S A **95**(13): 7445–50.

Wysocki, R., A. Javaheri, S. Allard, F. Sha, J. Cote and S. J. Kron (2005). "Role of Dot1-dependent histone H3 methylation in G1 and S phase DNA damage checkpoint functions of Rad9." Mol Cell Biol **25**(19): 8430–43.

Zhang, J., H. Willers, Z. Feng, J. C. Ghosh, S. Kim, D. T. Weaver, J. H. Chung, S. N. Powell and F. Xia (2004). "Chk2 phosphorylation of BRCA1 regulates DNA double-strand break repair." Mol Cell Biol **24**(2): 708–18.

Zou, L. and S. J. Elledge (2003). "Sensing DNA damage through ATRIP recognition of RPA-ssDNA complexes." Science **300**(5625): 1542–8.

Chapter 4
Short Abstracts – Session I

Cancer: A Malady of Genes

Inder M. Verma

The Salk Institute, P.O. Box 85800, San Diego, CA 92186, USA

The transformation of a seemingly normal cell into a malignant one is governed by an elaborate network of molecular interactions that resembles an intricate electrical circuit. Prominent nodes in this circuit include enhanced chromosomal instability, loss of control of orderly cell division, inability to repair damaged genetic material, a reduction in pro-grammed cell death and an increase in cellular life span, and augmentation of the blood supply to tumors (angiogenesis). In the last two decades, we have witnessed the growth of an unprecedented understanding of this molecular circuitry. The conservation of the basic cellular machinery from yeast to humans, organisms that evolved some billion years apart, gives us hope that we will succeed in dissecting the entire genetic network involved in the life of both normal and abnormal cells. Genetic or epigenetic events leading to mutations in genes or alterations in gene activity can lead either to the gene becoming an oncogene, where the protein it codes gains a new and deleterious function, or to the loss of the function of a protein essential to the regulation of cell turnover (tumor suppressors). Changes of either kind may induce neoplasia – an uncontrolled and disorderly proliferation of cells to form a tumor. In some simplistic way the problem of cancer seems disarmingly easy: stop the function of uncontrolled oncogenes and replenish the products of deficient tumor-suppressor genes.Critics say that despite the wealth of this molecular knowledge, the three principal modalities of treating cancer, that is, surgery, radiotherapy and chemotherapy, have remained essentially the main defence against cancer for the past 50 years. But in my view that is a rather unfair assessment of the knowledge we have garnered in understanding the nature and biology of malignant cells, on which the development of new therapeutics depends. The selective chemotherapeutic agents that are beginning to come into the clinic, therapeutic monoclonal antibodies, novel anti-angiogenesis molecules, inhibitors of the kinases that lead to uncontrolled cell proliferation and proteosomal inhibitors that prevent protein degradation, all have their direct origin in the knowledge gained from deciphering the molecular events involved in cell regulation. I will elaborate some of these concepts with work from our laboratory.

Understanding Transcription and Signalling in Cancer Epigenetics

Colin Goding

Marie Curie Research Institute, The Chart, Oxted, Surrey, RH8 OTL, UK

The ability of cancer cells to acquire properties of invasiveness and the potential to metastasise is not entirely understood. In the traditional model, the acquisition of cells with metastatic potential is acquired by the accumulation of genetic lesions. Alternatively, metastatic potential could represent a specific epigenetic state that may be inherently unstable; properties associated with metastasis could be lost once the metastatic cell has taken up residence in another location. However, while considerable resources have been expended on trying to pinpoint mutations that correlate with metastatic potential, the precise molecular mechanisms underlying any epigenetic model for cancer metastasis have been relatively little explored. In the melanocyte lineage and melanoma the Microphthalmia-associated transcription factor Mitf plays a crucial role in commitment, survival, and differentiation and regulates G1/S transition through up-regulation of the p16INK4a and p21Cip1-cyclin-dependent kinase inhibitors. We have now uncovered a mechanism in which Mitf promotes proliferation and inhibits invasiveness through regulation of the actin cytoskeleton. The results suggest that within the melanoma microenvironment, variations in the repertoire of signals that determine Mitf activity will dictate the differentiation, proliferative and invasive potential of individual cells through a dynamic epigenetic mechanism. Moreover, rather than operating as simple on–off switches, transcription factors like Mitf act to integrate a range of inputs to generate a variable biological output.

Matrix Metalloproteinase-Induced Malignancy

Derek Radisky

Mayo Clinic Cancer Center, Griffin Cancer Research Building 4500 San Pablo RoadJacksonville, FL 32224, USA

The tumor microenvironment can facilite cancer progression and activate dormant cancer cells, and also can stimulating tumor formation. Previous investigations of stromelysin-1/matrix metalloproteinase-3 (MMP-3), a stromal enzyme upregulated in many breast tumors, found that MMP-3 can cause epithelial–mesenchymal transition (EMT) and malignant transformation in cultured cells, and genomically unstable mammary carcinomas in transgenic mice3. We have elucidated the molecular pathways by which MMP-3 exerts these effects, and shown that exposure of mouse mammary epithelial cells to MMP-3 induces expression of Rac1b, an alternatively spliced isoform of Rac1, which causes an increase in cellular reactive oxygen species (ROS). The ROS stimulate expression of the transcription factor Snail and EMT, and cause oxidative DNA damage and genomic instability. These findings identify a novel pathway in which a component of the breast tumor microenvironment alters cellular structure in culture and tissue structure in vivo, leading to malignant transformation. Recently, we have extended these findings to show that other MMPs can also induce these effects, and that other organ systems are responsive to these pathways.

Signaling by the TOR Pathway to Akt PKB and S6K

David M. Sabatini

Whitehead Institute, Cambridge, MA, USA

mTOR is the target of the immunosuppressive drug rapamycin and the central component of a nutrient- and hormone-sensitive signaling pathway that regulates cell growth and proliferation. We have identified two distinct mTOR-containing proteins complexes, one of which regulates growth through S6K and another that regulates cell survival through Akt. These complexes define both rapamycin-sensitive and insensitive branches of the mTOR pathway. I will provide an overview of mTOR signaling as well as discuss the mechanism of action of rapamycin.

ATM Regulates ATR Chromatin Loading in Response to DNA Double-Strand Breaks

Myriam Cuadrado, Barbara Martinez, Matilde Murga and Oskar Fernandez-Capetillo

Genomic Instability Group, Spanish National Cancer Center. Madrid, Spain

DNA double-strand breaks (DSBs) are among the most deleterious lesions that can challenge genomic integrity. Concomitant to the repair of the breaks, a rapid signaling cascade must be coordinated at the lesion site that leads to the activation of cell cycle checkpoints and/or apoptosis. In this context, ataxia telangiectasia mutated (ATM) and ATM and Rad-3-related (ATR) protein kinases are the earliest signaling molecules that are known to initiate the transduction cascade at damage sites. The current model places ATM and ATR in separate molecular routes that orchestrate distinct pathways of the checkpoint responses. Whereas ATM signals DSBs arising from ionizing radiation (IR) through a Chk2-dependent pathway, ATR is activated in a variety of replication-linked DSBs and leads to activation of the checkpoints in a Chk1 kinase-dependent manner. However, activation of the G2/M checkpoint in response to IR escapes this accepted paradigm because it is dependent on both ATM and ATR but independent of Chk2. Our data provides an explanation for this observation and places ATM activity upstream of ATR recruitment to IR-damaged chromatin. These data provide experimental evidence of an active cross talk between ATM and ATR signaling pathways in response to DNA damage.

BRAD1 is the Founding Member of a New Family Protein in pRB/E2F Pathway, Amplified and Overexpressed Aggressive Tumors, with Chromatin Remodeling Activity Cooperating MYC Oncogene

Marco Ciro[1], Elena Prosperini[1], Ursula Grazini[1], Micaela Quarto[2], Maria Capra[2], Giovanni Pacchiana[1], Fraser McBlane[1] and Kristian Helin[1-3]

[1]*Department of Experimental Oncology, European Institute of Oncology, Milan, Italy*
[2]*FIRC Institue of Molecular Oncology, Milan, Italy*
[3]*Biotech Research*

Most human tumors contain genetic alterations in the pRB/E2F pathway. The identification of genes downstream of the pRB/E2F pathway and overexpressed in tumors will gain novel insights into how the E2Fs regulate cell proliferation and will lead to the identification of new critical players in human cancer. Based on this approach, we have identified a new gene named *BRomodomain and ATPase Domain 1* (*BRAD1*) after its domain composition. Consistent with being a direct E2F target, *BRAD1* gene is cell growth and cell cycle regulated, accumulating in S-phase. *BRAD1* is also required for proper G0 to G1/S transition and for efficient colony forming ability of both normal and cancer cells. BRAD1 is also overexpressed and amplified in a significant number of human cancers, correlating with tumor grade III. *BRAD1* is associated with chromatin and nuclear matrix. In agreement with the presence of an ATPase domain and a Bromodomain, we have shown that *BRAD1* containing complex alters the accessibility of polynucleosomal arrays. Very interestingly, we provide evidence that *BRAD1* interacts with the *MYC* oncogene, binds to genomic targets of *MYC* and cooperates with *MYC* dependent function. Based on this data, we propose that *BRAD1* is the founding member of a new family of gene in the pRB/E2F pathway, amplified and overexpressed in aggressive tumors, associated to a chromatin remodeling activity and cooperating with *MYC* oncogene.

Inflammatory Cytokines and Cancer

Frances Balkwill

Centre for Translational Oncology and Cancer Research UK Clinical Centre Barts and The London, Queen Marys' Medical School, London EC1Y 2AN, UK

Neoplastic tissue selects for the type and extent of inflammation most favourable to tumor growth and progression. Several lines of evidence, including general or cell-specific gene inactivation, population-based studies and genetic analysis, are consistent with the view that inflammatory cytokines play an important role in malignant progression. These cytokines promote cancer cell survival and proliferation, aid metastasis, stimulate stroma deposition and remodelling, skew and tame adaptive immunity, and enhance angiogenesis. One example of this is the inflammatory cytokine TNF-α that can act as an endogenous tumor promoter in experimental cancers of the skin, liver, colon and ovary. In these models, and in some human cancers, low levels of TNF-α can initiate and sustain production of other cytokines and chemokines, as well as matrix metalloproteases in both tumor cells and associated stroma. If TNF-α production becomes chronic and deregulated, the tumor continues to make these inflammatory mediators and these attract leucocytes and other stromal cells. Stromal cells provide growth and survival factors, contribute to ECM remodelling and development of new blood vessels, and create an immunosuppressive microenvironment. Stromal and malignant cells communicate via autocrine and paracrine loops, generally enhanced by positive feedback. The malignant cells proliferate and invade locally. Some of the malignant cells acquire chemokine receptors, and this expression can be enhanced by TNF-α. Chemokine receptor-expressing cells are able to respond to chemokine gradients at sites distant from the primary tumor. Chemokines produced at sites of metastasis induce a pulse of TNF-α that can stimulate once again a tumor-promoting inflammatory network, encouraging proliferation and survival of the metastatic cells in their new location and development of a supporting stroma. Therapeutic targeting of cancer-associated inflammatory cytokines is in its infancy, but initial clinical results are encouraging, and may complement more conventional strategies.

Section 2
Tumor and Microenvironment Interactions

Chapter 5
Tumor Promotion by Tumor-Associated Macrophages

Chiara Porta[1], Biswas Subhra Kumar[2], Paola Larghi[1], Luca Rubino[1], Alessandra Mancino[1] and Antonio Sica[1, 2]*

[1]*Fondazione Humanitas per la Ricerca, Istituto Clinico Humanitas, 20089 Milan, Italy*
[2]*Institute of Molecular and Cell Biology, Biopolis Drive (Proteos), Singapore*

Address correspondence to: Antonio Sica, Istituto Clinico Humanitas, 20089 Milan, Italy; Tel: +39.02.8224.5111; fax: +39.02.8224.5101
*e-mail: antonio.sica@humanitas.it

Abstract. Recent years have seen a renaissance of the inflammation-cancer connection stemming from different lines of work and leading to a generally accepted paradigm (Balkwill and Mantovani 2001; Mantovani et al. 2002; Coussens and Werb 2002; Balkwill et al. 2005).

An inflammatory component is present in the microenvironment of most neoplastic tissues, including those not causally related to an obvious inflammatory process. Cancer-associated inflammation includes: the infiltration of white blood cells, prominently phagocytic cells called macrophages (TAM) (Paik et al. 2004); the presence of polipeptide messengers of inflammation (cytokines such as tumor necrosis factor (TNF) or interleukin-1 (IL-1), chemokines such as CCL2); the occurrence of tissue remodelling and angiogenesis.

Chemokines have emerged as a key component of the tumor microenvironment which shape leukocyte recruitment and function (Pollard 2004). Strong direct evidence suggests that cancer associated inflammation promotes tumor growth and progression. Therapeutic targeting of cancer promoting inflammatory reactions is in its infancy, and its development is crucially dependent on defining the underlying cellular and molecular mechanisms in relevant systems.

1 Introduction

Recent years have seen a renaissance of the inflammation-cancer connection stemming from different lines of work and leading to a generally accepted paradigm (Balkwill et al. 2001; Coussens et al. 2002; Balkwill et al. 2005). Epidemiological studies have revealed that chronic inflammation predisposes to different forms of cancer, such as colon, prostate, and liver cancer, and that usage of nonsteroidal anti-inflammatory agents can protect against the emergence of various tumors. An inflammatory component is present in the microenvironment of most neoplastic tissues, including those not causally related to an obvious inflammatory process. Hallmarks of cancer-associated inflammation include the infiltration of white blood cells, the presence of polypeptide messengers of inflammation (cytokines and chemokines), and the occurrence of tissue remodeling and angiogenesis.

Strong evidence suggests that cancer-associated inflammation promotes tumor growth and progression (Balkwill et al. 2001; Coussens et al. 2002; Balkwill et al. 2005). By late 1970s it was found that tumor growth is promoted by tumor-associated macrophages (TAM), a major leukocyte population present in tumors (Balkwill et al. 2001; Coussens et al. 2002; Balkwill et al. 2005; Mantovani et al. 2002). Accordingly, in many but not all human tumors, a high frequency of infiltrating TAM is associated with poor prognosis. Interestingly, this pathological finding has reemerged in the postgenomic era: genes associated with leukocyte or macrophage infiltration (e.g., CD68) are part of the molecular signatures which herald poor prognosis in lymphomas and breast carcinomas (Paik et al. 2004). Gene-modified mice and cell transfers have provided direct evidence for the pro-tumor function of myeloid cells and their effector molecules. These results raise the interesting possibility of targeting myelomonocytic cells associated with cancer as an innovative therapeutic strategy. Here, we will review key properties of TAM emphasizing recent genetic evidence and emerging targets for therapeutic intervention.

2 Macrophage Polarization

Heterogeneity and plasticity are hallmarks of cells belonging to the monocyte-macrophage lineage (Gordon 2003; Mantovani et al. 2002; Mantovani, Sica, Sozzani, Allavena, Vecchi, and Locati 2004). Lineage-defined populations of mononuclear phagocytes have not been identified but, already at the short-lived stage of circulating precursor, monocyte subsets characterized by differential expression of the FcγRIII receptor (CD16) or of chemokine receptors (CCR2, CX3CR1, and CCR8) and by different functional properties have been described. Once in tissues, macrophages acquire distinct morphological and functional properties directed by the tissue (e.g., the lung alveolar macrophage) and immunological microenvironment.

In response to cytokines and microbial products, mononuclear phagocytes express specialized and polarized functional properties (Gordon 2003; Mantovani et al. 2004, Ghassabech, et al. 2006). Mirroring the Th1/Th2 nomenclature, many refer to polarized macrophages as M1 and M2 cells. Classically activated M1 macrophages have long been known to be induced by IFNγ alone or in concert with microbial stimuli (e.g., LPS) or cytokines (e.g., TNF and GM-CSF). IL-4 and IL-13 were subsequently found to be more than simple inhibitors of macrophage activation and to induce an alternative M2 form of macrophage activation (Gordon 2003). M2 is a generic name for various forms of macrophage activation other than the classic M1, including cells exposed to IL-4 or IL-13, immune complexes, IL-10, glucocorticoid, or secosteroid hormones (Mantovani et al 2004).

In general, M1 cells have an IL-12^{high}, IL-23^{high}, IL-10^{low} phenotype; are efficient producers of effector molecules (reactive oxygen and nitrogen intermediates) and inflammatory cytokines (IL-1β, TNF, IL-6); participate as inducer and effector cells in polarized Th1 responses; mediate resistance against intracellular parasites and tumors. In contrast, the various forms of M2 macrophages share an IL-12^{low}, IL-23^{low}, IL-10^{high} phenotype with variable capacity to produce inflammatory cytokines depending on the signal utilized. M2 cells generally have high levels of scavenger, mannose and galactose-type receptors, and the arginine metabolism is shifted to ornithine and polyamines. Differential regulation of components of the IL-1 system (Dinarello 2005) occurs in polarized macrophages, with low IL-1β and low caspase I, high IL-1ra and high decoy type II receptor in M2 cells. M1 and the various forms of M2 cells have distinct chemokine and chemokine receptor repertoires (Mantovani et al. 2004). In general, M2 cells participate in polarized Th2 reactions, promote killing and encapsulation of parasites (Noel et al. 2004), are present in established tumors and promote progression, tissue repair and remodeling (Wynn 2004), and have immunoregulatory functions (Mantovani et al. 2004). Immature myeloid suppressor cells have functional properties and a transcriptional profile related to M2 cells (Biswas et al. 2006). Moreover, polarization of neutrophil functions has also been reported (Tsuda et al. 2004).

Profiling techniques and genetic approaches have put to a test and shed new light on the M1/M2 paradigm. Transcriptional profiling has offered a comprehensive picture of the genetic programmes activated in polarized macrophages, led to the discovery of new polarization-associated genes (e.g., *Fizz* and *YM-1*), tested the validity of the paradigm in vivo in selected diseases (Takahashi et al. 2004; Desnues et al. 2005; Biswas et al. 2006), and questioned the generality of some assumptions. For instance, unexpectedly arginase is not expressed prominently in human IL-4-induced M2 cells (Scotton et al. 2005). M2 cells express high levels of the chitinase-like YM-1. Chitinases represent an antiparasite strategy conserved in evolution and there is now evidence that acidic mammalian chitinase induced by IL-13 in macrophages is an important mediator of type II inflammation (Zhu et al. 2004).

3 Macrophage Recruitment at the Tumor Site

TAM derive from circulating monocytes and are recruited at the tumor site by a tumor-derived chemotactic factor (TDCF) for monocytes, originally described by this group (Bottazzi et al. 1983) and later identified as the chemokine CCL2/MCP-1 (Yoshimura et al. 1989; Matsushima et al. 1999) (Figure 1). Evidence supporting a pivotal role of chemokines in the recruitment of monocytes in neoplastic tissues includes: correlation between

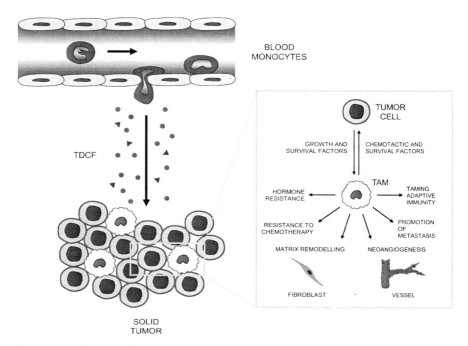

FIGURE 1. Tumor-derived chemotactic factors (CC-chemokines, e.g., CCL2, Macrophage colony stimulating factor (M-CSF) and vascular endothelial growth factor (VEGF), actively recruit circulating blood monocytes at the tumor site. In the tumor microenvironment monocytes differentiate into tumor-associated macrophages (TAM) which establish a symbiotic relationship with tumor cells. The above tumor-derived factors positively modulate TAM survival. From their own, TAM secrete growth factors which promote tumor cell proliferation and survival; regulate matrix deposition and remodeling and activates neo-angiogenesis.

production and infiltration in murine and human tumors, passive immunization and gene modification (Rollins 1999). In addition, the central role of chemokines in shaping the tumor microenvironment is supported by the observation that tumors are generally characterized by the constitutive expression of chemokines belonging to the inducible realm (Mantovani 1999). The molecular mechanisms accounting for the constitutive expression of chemokines by cancer cells have been defined only for CXCL1 and involve NF-κB activation by NF-κB-inducing kinase (Yang and Richmond 2001).

CCL2 is probably the most frequently found CC chemokine in tumors. Most human carcinomas produce CCL2 (Table 1) and its levels of expression correlate with the increased infiltration of macrophages (Mantovani et al. 2002; Balkwill 2004; Conti and Rollins 2004). Interestingly, CCL2 production has also been detected in TAM, indicating the existence of an amplification loop for their recruitment (Mantovani et al. 2002; Ueno et al. 2000). Other

CC chemokines related to CCL2, such as CCL7 and CCL8, are also produced by tumors and shown to recruit monocytes (Van Damme et al. 1992).

Along with the supposed protumoral role of TAM, the local production of chemokines and the extent of TAM infiltration have been studied as prognostic factors. For example, in human breast and oesophagus cancers, CCL2 levels correlated with the extent of macrophage infiltration, lymph-node metastasis and clinical aggressiveness (Azenshtein et al. 2002; Saji et al. 2001). In an experimental model of nontumorigenic melanoma, low-level of CCL2 secretion, with "physiological" accumulation of TAM, promoted tumor formation, while high CCL2 secretion resulted in massive macrophage infiltration into the tumor mass and in its destruction (Nesbit et al. 2001). In pancreatic cancer patients, high serum levels of CCL2 were associated with more favourable prognosis and with a lower proliferative index of tumor cells (Monti et al. 2003). These biphasic effects of CCL2 are consistent with the "macrophage balance" hypothesis (Mantovani et al. 1992) and emphasise the concept that levels of macrophage infiltration similar to those observed in human malignant lesions express protumor activity (Bingle et al. 2000).

A variety of other chemokines have been detected in neoplastic tissues as products of either tumor cells or stromal elements (Table 1). These molecules play an important role in tumor progression by direct stimulation of neoplastic

TABLE 1. Tumor-derived chemokines.

Ligand	Producing tumor
CXC family	
CXCL1/GroA	Colon carcinoma (Li et al. 2004)
CXCL8/IL8	Melanoma (Haghnegahdar et al. 2000), breast (Azenshtein et al. 2005)
CXCL9/Mig	Hodgkin's disease (Teruya-Feldstein, Tosato, and Jaffe 2000)
CXCL10/IP-10	Hodgkin lymphoma and nasopharyngeal carcinoma (Teichmann et al. 2005)
CXCL12/SDF-1	Melanoma (Scala et al. 2005)
CXCL13/BCA1	Non-Hodgkin B-cell lymphoma (Smith et al. 2003)
CC family	
CCL1/I-309	Adult T-cell leucemia (Ruckes et al. 2001)
CCL2/MCP1	Pancreas (Saji et al. 2001); sarcomas, gliomas, lung, breast, cervix, ovary, melanoma (Mantovani 1999)
CCL3/MIP-1a	Schwann cell tumors (Mori et al. 2004)
CCL3LI/LD78b	Glioblastoma (Kouno et al. 2004)
CCL5/RANTES	Breast (Ueno et al. 2000); melanoma (Payne and Cornelius 2002)
CCL6	NSLC (Yi et al. 2003)
CCL7/MCP-3	Osteosarcoma (Conti et al. 2004)
CCL8/MCP-2	Osteosarcoma (Conti et al. 2004)
CCL11/eotaxin	T-cell lymphoma (Kleinhans et al. 2003)
CCL17/TARC	Lymphoma (Vermeer et al. 2002)
CCL18/PARC	Ovary (Schutyser et al. 2002)
CCL22/MDC	Ovary (Sakaguchi 2005)
CCL28/MEC	Hodgkin's disease (Hanamoto et al. 2004)

growth, promotion of inflammation and induction of angiogenesis. In spite of constitutive production of neutrophil chemotactic proteins by tumor cells, CXCL8 and related chemokines, neutrophils are not a major and obvious constituent of the leukocyte infiltrate. However, these cells, though present in minute numbers, may play a key role in triggering and sustaining the inflammatory cascade. Macrophages are also recruited by molecules other than chemokines. In particular, tumor-derived cytokines interacting with tyrosine kinase receptors, such as vascular endothelial growth factor (VEGF) and macrophage colony stimulating factor (M-CSF) (Lin et al. 2001; Duyndam et al. 2002) promote macrophage recruitment, as well as macrophage survival and proliferation, the latter generally limited to murine TAM (Mantovani et al. 2002; Lin et al. 2001; Duyndam et al. 2002) (Figure 1). Using genetic approaches, it has been demonstrated that depletion of M-CSF markedly decreases the infiltration of macrophages at the tumor site, and this correlates with a significant delay in tumor progression. By contrast, over-expression of M-CSF by tumor cells dramatically increased macrophage recruitment and this was correlated with accelerated tumor growth (Lin et al. 2001; Nowicki et al. 1996; Aharinejad et al. 2002). M-CSF over-expression is common among tumors of the reproductive system, including ovarian, uterine, breast and prostate, and correlates with poor prognosis (Pollard 2004). Recently, placenta-derived growth factor (PlGF), a molecule related to VEGF in terms of structure and receptor usage, has been reported to promote the survival of TAM (Adini et al. 2002).

4 TAM Express Selected M2 Protumoral Function

The cytokine network expressed at the tumor site plays a central role in the orientation and differentiation of recruited mononuclear phagocytes, thus contributing to direct the local immune system away from antitumor functions (Mantovani et al. 2002). This idea is supported by both preclinical and clinical observations (Bingle et al. 2000; Goerdt and Orfanos 1999) that clearly demonstrate an association between macrophage number/density and prognosis in a variety of murine and human malignancies. The immunosuppressive cytokines IL-10 and TGFβ are produced by both cancer cells (ovary) and TAM (Mantovani et al. 2002). IL-10 promotes the differentiation of monocytes to mature macrophages and blocks their differentiation to dendritic cells (DC) (Allavena et al. 2000) (Figure 2). Thus, a gradient of tumor-derived IL-10 may account for differentiation along the DC versus the macrophage pathway in different microanatomical localizations in a tumor. Such situation was observed in papillary carcinoma of the thyroid, where TAM are evenly distributed throughout the tissue, in contrast to DC which are present in the periphery (Scarpino et al. 2000). In breast carcinoma, DC with a mature phenotype (DC-LAMP+) were localized in peritumoral areas, while immature DC were inside the tumor (Bell et al. 1999). Interestingly, it was shown that

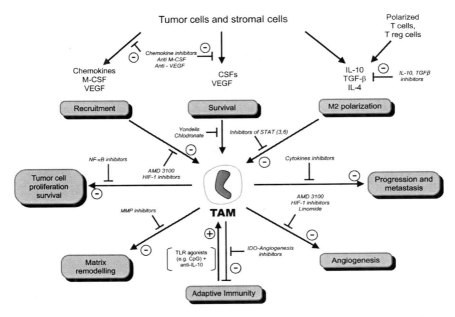

FIGURE 2. Multifaced therapeutic approaches to prevent TAM protumoral functions. TAM promote tumor progression by favoring angiogenesis, suppression of adaptive immunity, matrix remodeling, tumor progression, and metastasis. The figure summarizes strategies impairing selective TAM protumoral functions (−) or restoring antitumor activities (+). Cytotoxic drugs (e.g., Yondelis) may decrease TAM number and prevent protumoral functions. A similar result may be obtained by limiting TAM recruitment (Linomide, HIF-1 inhibitors, AMD3100). Restoration of M1 immunity (STAT-3 and −6 inhibitors; anti-IL-10 plus CpG; IDO inhibitors) would provide cytotoxic activity and reactivation of Th1 specific antitumor immunity. Inhibition of both proinflammatory cytokines and growth factors expression (NF-κB inhibitors) may disrupt inflammatory circuits supporting tumor growth and progression. MMPs inhibitors would prevent cancer cells spread and metastasis. Finally, inhibitors of TAM-mediated angiogenesis (Linomide, HIF-1 inhibitors, AMD3100) would restrain blood supply and inhibit tumor growth. Macrophage-colony stimulating factor (M-CSF); VEGF (Vascular Endothelial Growth Factor); CSFs (Colony-Stimulating Factors); IL- (Interleukin-); TGF-β (Transforming Growth Factor-β); IDO (indoleamine 2,3-dioxygenase); MMP inhibitors (Matrix metalloproteinase inhibitors); TLR agonists (Toll-Like Receptor agonists); STAT (signal transducer and activator of transcription); NF-κB (nuclear factor-kappaB).

Stat3 is constitutively activated in tumor cells (Wang et al. 2004) and in diverse tumor-infiltrating immune cells (Kortylewski et al. 2005), leading to inhibition of the production of several proinflammatory cytokines and chemokines and to the release of factors that suppress dendritic cell maturation. Noteworthy, ablating Stat3 in hematopoietic cells triggers an intrinsic immune-surveillance system that inhibits tumor growth and metastasis (Kortylewski et al. 2005).

As previously discussed, IL-10 promotes the M2c alternative pathway of macrophage activation and induce TAM to express M2-related functions. Indeed, under many aspects TAM summarize a number of functions expressed by M2 macrophages, involved in tuning inflammatory responses and adaptive immunity, scavenge debris, promote angiogenesis, tissue remodeling, and repair. The production of IL-10, TGFβ, and PGE2 by cancer cells and TAM (Mantovani et al. 2002) contributes to a general suppression of antitumor activities (Figure 2).

TAM are poor producers of NO (DiNapoli et al. 1996) and, in situ in ovarian cancer, only a minority of tumors and, in these, a minority of macrophages localized at the periphery scored positive for iNOS (Klimp et al. 2001). Moreover, in contrast to M1 polarized macrophages, TAM have been shown to be poor producers of reactive oxygen intermediates (ROIs), consistent with the hypothesis that these cells represent a skewed M2 population (Klimp et al. 2001).

Moreover, TAM were reported to express low levels of inflammatory cytokines (e.g., IL-12, IL-10, TNFα, IL-6) (Mantovani et al. 2002). Activation of NF-κB is a necessary event promoting transcription of several proinflammatory genes. Our previous studies (Sica et al. 2000) indicated that TAM display defective NF-κB activation in response to the M1 polarizing signal LPS and we observed similar results in response to the proinflammatory cytokines TNFα and IL-1β (Biswas et al. 2006). Thus in terms of cytotoxicity and expression of inflammatory cytokines TAM resemble the M2 macrophages. Unexpectedly, TAM display high level of IRF-3/STAT-1 activation, which may be part of the molecular events promoting TAM-mediated T cell deletion (Kusmartsev and Gabrilovich 2005).

In agreement with the M2 signature, TAM also express high levels of both the scavenger receptor-A (SR-A) (Biswas et al. 2006) and the mannose receptor (MR) (Allavena, unpublished observation). Further, TAM are poor antigen presenting cells (Mantovani et al. 2002).

Arginase expression in TAM has not been studied. However, it has been recently proposed that the carbohydrate-binding protein galectin-1, which is abundantly expressed by ovarian cancer (Van Den Brule, Califice, Garnier, Fernandez, Berchuck, and Castronovo 2003) and shows specific antiinflammatory effects, tunes the classic pathway of L-arginine resulting in a strong inhibition of the nitric oxide production by lipopolysaccharide-activated macrophages.

Angiogenesis is an M2-associated function which represents a key event in tumor growth and progression. In several studies in human cancer, TAM accumulation has been associated with angiogenesis and with the production of angiogenic factors such as VEGF and platelet-derived endothelial cell growth factor (Mantovani et al. 2002). More recently, in human cervical cancer, VEGF-C production by TAM was proposed to play a role in peritumoral lympho-angiogenesis and subsequent dissemination of cancer cells with formation of lymphatic metastasis (Schoppmann et al. 2002).

Additionally, TAM participate to the proangiogenic process by producing the angiogenic factor thymidine phosporylase (TP), which promotes endothelial cell migration in vitro and whose levels of expression are associated with tumor neovascularization (Hotchkiss et al. 2003). TAM contribute to tumor progression also by producing proangiogenic and tumor-inducing chemokines, such as CCL2 (Vicari and Caux 2002). Moreover, TAM accumulate in hypoxic regions of tumors and hypoxia triggers a proangiogenic program in these cells (see later). Therefore, macrophages recruited in situ represent an indirect pathway of amplification of angiogenesis, in concert with angiogenic molecules directly produced by tumor cells. On the antiangiogenic side, in a murine model, GM-CSF released from a primary tumor up-regulated TAM-derived metalloelastase and angiostatin production, thus suppressing tumor growth of metastases (Dong et al. 1998).

Finally, TAM express molecules which affect tumor cell proliferation, angiogenesis and dissolution of connective tissues. These include epidermal growth factor (EGF), members of the FGF family, TGFβ, VEGF, and chemokines. In lung cancer, TAM may favor tumor progression by contributing to stroma formation and angiogenesis through their release of PDGF, in conjunction with TGF-β1 production by cancer cells (Mantovani et al. 2002). Macrophages can produce enzymes and inhibitors which regulate the digestion of the extracellular matrix, such as MMPs, plasmin, urokinase-type plasminogen activator (uPA) and the uPA receptor. Direct evidence have been presented that MMP-9 derived from hematopoietic cells of host origin contributes to skin carcinogenesis (Coussens et al. 2000). Chemokines have been shown to induce gene expression of various MMPs and, in particular, MMP-9 production, along with the uPA receptor (Locati et al. 2002). Evidence suggests that MMP-9 has complex effects beyond matrix degradation including promotion of the angiogenesis switch and release of growth factors (Coussens et al. 2000).

The mechanisms responsible for the M2 polarization of TAM have not been completely defined yet. Recent data point to tumor (ovarian, pancreatic) derived signals which promote M2 differentiation of mononuclear phagocytes (Hagemann et al. unpublished data).

5 Modulation of Adaptive Immunity by TAM

It has long been known that TAM have poor antigen-presenting capacity and can actually suppress T cell activation and proliferation (Mantovani et al. 2002). The suppressive mediators produced by TAM include prostaglandins, IL-10 and TGFβ and indoleamine dioxigenase (IDO) metabolites (Mantovani et al. 2004). Moreover, TAM are unable to produce IL-12, even upon stimulation by IFN-γ and LPS (Sica et al. 2000). With this cytokine profile, which is characteristic of M2 macrophages, TAM are unable to trigger Th1 polarized immune responses, but rather induce T regulatory cells (Treg). Treg cells possess a characteristic anergic phenotype and strongly suppress the activity of effector T cells and other inflammatory cells, such as monocytes. Infiltrating Treg cells strongly affect the tumor microenvironment by producing high level of immunosuppressive cytokines (IL-10, TGFβ) (Jarnicki et al. 2006). Suppression of T-cell mediated antitumor activity by Treg cells is associated with increased tumor growth and, hence, decreased survival (Sakaguchi 2005). For instance, in patients with advanced ovarian cancer, an increase in the number of functionally active Treg cells present in the ascites was predictive of reduced survival (Curiel et al. 2004). Immature myeloid suppressor cells present in the neoplastic tissue of some tumors have been shown to potently inhibit T cell responses (Bronte et al. 2003). The relationship, if any, of immature myeloid suppressor cells with TAM remains to be defined.

The complex network of chemokines present at the tumor site can play a role also in the induction of the adaptive immunity. Chemokines also regulate the amplification of polarized T cell responses. Some chemokines may enhance specific host immunity against tumors but on the other hand other chemokines may contribute to escape from the immune system, by recruiting Th2 effectors and Treg cells (Mantovani et al. 2004). As mentioned above, in addition to being a target for chemokines, TAM are a source of a selected set of these mediators (CCL2, CCL17, CCL18, CCL22). CCL18 was recently identified as the most abundant chemokine in human ovarian ascites fluid. When the source of CCL18 was investigated, it was tracked to TAM, with no production by ovarian carcinoma cells (Schutyser et al. 2002). CCL18 is a CC chemokine produced constitutively by immature DC and inducible in macrophages by IL-4, IL-13, and IL-10. Since IL-4 and IL-13 are not expressed in substantial amounts in ovarian cancer, it is likely that IL-10, produced by tumor cells and macrophages themselves, accounts for CCL18 production by TAM. CCL18 is an attractant for naive T cells by interacting with an unidentified receptor (Adema et al. 1997). Attraction of naive T cells in a peripheral microenvironment dominated by M2 macrophages and immature DC is likely to induce T cell anergy.

Work in gene-modified mice has shown that CCL2 can orient specific immunity in a Th2 direction. Although the exact mechanism for this action has not been defined, it may include stimulation of IL-10 production in macrophages (Gu et al. 2000). Overall, TAM-derived chemokines most frequently recruit effector T cell inefficient to mount a protective anti-tumor immunity. TAM also produce chemokine specifically attracting T cells with immunosuppressive functions.

6 TAM as a Therapeutic Target

Major functions of TAM potentially amenable of therapeutic interventions are their activation, recruitment, angiogenic activity, survival, matrix remodeling and immunosuppression.

Defective NF-κB activation in TAM correlates with impaired expression of NF-κB-dependent inflammatory functions (e.g., expression of cytotoxic mediators, NO) and cytokines (TNFα, IL-1, IL-12) (Mantovani et al. 2002; Sica et al. 2000; Torroella-Kouri et al. 2005). Restoration of NF-κB activity in TAM is therefore a potential strategy to restore M1 inflammation and intra-tumoral cytotoxicity. In agreement, recent evidence indicates that restoration of an M1 phenotype in TAM may provide therapeutic benefit in tumor bearing mice. In particular, combination of CpG plus an anti-IL-10 receptor antibody switched infiltrating macrophages from M2 to M1 and triggered innate response debulking large tumors within 16 h (Guiducci et al. 2005). It is likely that this treatment may restore NF-κB activation and inflammatory functions by TAM. Moreover, TAM from STAT6 −/− tumor bearing mice display an M1 phenotype, with low level of arginase and high level of NO. As a result, these mice immunologically rejected spontaneous mammary carcinoma (Sinha et al. 2005). These data suggest that switching the TAM phenotype from M2 to M1 during tumor progression may promote antitumor activities. In this regard, the SHIP1 phosphatase was shown to play a critical role in programming macrophage M1 versus M2 functions. Mice deficient for SHIP1 display a skewed development away from M1 macrophages (which have high inducible nitric oxide synthase levels and produce NO), towards M2 macrophages (which have high arginase levels and produce ornithine) (Rauh et al. 2004). Finally, recent reports have identified a myeloid M2-biased cell population in lymphoid organs and peripheral tissues of tumor-bearing hosts, referred to as the myeloid suppressor cells (MSC), which are suggested to contribute to the immunosuppressive phenotype (Bronte et al. 2005). These cells are phenotypically distinct from TAM and are characterized by the expression of the Gr-1 and CD11b markers. MSC use two enzymes involved in the arginine metabolism to control T cell response: inducible nitric oxide synthase (NOS2) and arginase (Arg1), which deplete the milieau of arginine, causing peroxinitrite generation, as well as lack of CD3ζ chain expression and T cell apoptosis. In prostate cancer, selective antagonists of these two enzymes were proved beneficial in restoring T cell-mediated cytotoxicity (Bronte et al. 2005).

Chemokines and chemokine receptors are a prime target for the development of innovative therapeutic strategies in the control of inflammatory disorders. Recent results suggest that chemokine inhibitors could affect tumor growth by reducing macrophage infiltration. Preliminary results in *MCP-1/CCL2* gene targeted mice suggest that this chemokine can indeed promote progression in a Her2/neu-driven spontaneous mammary carcinoma model (Conti et al. 2004). Thus, available information suggests that chemokines represent a valuable therapeutic target in neoplasia. CSF-1 was identified as an important regulator of mammary tumor progression to metastasis, by regulating infiltration and function of TAM. Transgenic expression of CSF-1 in mammary epithelium led to the acceleration of the late stages of carcinoma and increased lung metastasis, suggesting that agents directed at CSF-1/CSF-1R activity could have important therapeutic effects (Pollard 2004; Aharinejad et al. 2002). Recent results have shed new light on the links between certain TAM chemokines and genetic events that cause cancer. The CXCR4 receptor lies downstream of the vonHippel/Lindau/hypoxia inducible factor (HIF) axis. Transfer of activated ras into a cervical carcinoma line, HeLa, induces IL-8/CXCL8 production that is sufficient to promote angiogenesis and progression. Moreover, a frequent early and sufficient gene rearrangement that causes papillary thyroid carcinoma (Ret-PTC) activates an inflammatory genetic program that includes CXCR4 and inflammatory chemokines in primary human thyrocytes (Borrello et al. 2005). The emerging direct connections between oncogenes, inflammatory mediators and the chemokine system provide a strong impetus for exploration of the anticancer potential of anti-inflammatory strategies. It was further demonstrated in non-small cell lung cancer (NSCLC) that mutation of the tumor suppressor gene *PTEN* results in up-regulation of HIF-1 activity and ultimately in HIF-1-dependent transcription of the *CXCR4* gene, which provides a mechanistic basis for the up-regulation of *CXCR4* expression and promotion of metastasis formation (Phillips et al. 2005). It appears therefore that targeting HIF-1 activity may disrupt the HIF-1/CXCR4 pathway and affect TAM accumulation, as well as cancer cell spreading and survival. The HIF-1-inducible VEGF is commonly produced by tumors and elicit monocyte migration. There is evidence that VEGF can significantly contribute to macrophage recruitment in tumors. Along with CSF-1, this molecule also promotes macrophage survival and proliferation.

VEGF is a potent angiogenic factor as well as a monocyte attractant that contributes to TAM recruitment. TAM promote angiogenesis and there is evidence that inhibition of TAM recruitment plays an important role in anti-angiogenic strategies. Linomide, an antiangiogenic agent, caused significant reduction of the tumor volume, in a murine prostate cancer model, by inhibiting the stimulatory effects of TAM on tumor angiogenesis (Joseph and Isaacs 1998). Based on this, the effects of Linomide, or other antiangiogenic drugs, on the expression of pro- and antiangiogenic molecules by TAM may be considered valuable targets for anticancer therapy. Due to the localization of TAM into the hypoxic regions of tumors, viral vectors were used to transduce macrophages with therapeutic genes, such as IFNγ, that were activated only in low oxygen conditions (Carta et al. 2001). These works present promising approaches which use macrophages as vehicles to deliver gene therapy in regions of tumor hypoxia.

Antitumor agents with selective cytotoxic activity on monocyte-macrophages would be ideal therapeutic tools for their combined action on tumor cells and TAM. We recently reported that Yondelis (Trabectedin), a natural product derived from the marine organism Ecteinascidia turbinata with potent anti-tumor activity (Sessa et al. 2005), is specifically cytotoxic to macrophages and TAM, while sparing the lymphocyte subset. This compound inhibits NF–Y, a transcription factor of major importance for mononuclear phagocyte differentiation. In addition, Yondelis inhibits the production of CCL2 and IL-6 both by TAM and tumor cells (Allavena et al. 2005). These antiinflammatory properties of Yondelis may be an extended mechanism of its antitumor activity.

TAM produce several matrix-metalloproteases (e.g., MMP2, MMP9) which degrade proteins of the extra-cellular matrix, and also produce activators of MMPs, such as chemokines (de Visser et al. 2006). Inhibition of this molecular pattern may prevent degradation of extracellular matrix, as well as tumor cell invasion and migration. The biphosphonate zoledronic acid is a prototipical MMP inhibitor. In cervical cancer this compound suppressed MMP-9 expression by infiltrating macrophages and inhibited metalloprotease activity, reducing angiogenesis and cervical carcinogenesis (Giraudo et al. 2004).The halogenated bisphosphonate derivative chlodronate is a macrophage toxin which depletes selected macrophage populations. Given the current clinical usage of this and similar agents it will be important to assess whether they have potential as TAM toxins.

The secreted protein acidic and rich in cysteine (SPARC) has gained much interest in cancer, being either up-regulated or down-regulated in progressing tumors. SPARC produced by macrophages present in tumor stroma can modulate collagen density, leukocyte, and blood vessel infiltration (Sangaletti et al. 2003).

Cyclooxygenase (COX) is a key enzyme in the prostanoid biosynthetic pathway. COX-2 is up-regulated by activated oncogenes (i.e., β-catenin, MET) but is also produced by TAM in response to tumor-derived factors like mucin in the case of colon cancer. The usage of COX-2 inhibitors in the form of nonsteroidal antiinflammatory drugs is associated with reduced risk of diverse tumors (colorectal, esophagus, lung, stomach, and ovary). Selective COX-2 inhibitors are now thought as part of combination therapy.

The IFN-γ-inducible enzyme indoleamine 2,3-dioxygenase is a well-known suppressor of T cell activation. It catalyzes the initial rate-limiting step in tryptophan catabolism, which leads to the biosynthesis of nicotinamide adenine dinucleotide. By depleting tryptophan from local microenvironment, IDO blocks activation of T lymphocytes. Recently, it was shown that inhibition of IDO may cooperate with cytotoxic agents to elicit regression of established tumors (Uyttenhove et al. 2003) and may increase the efficacy of cancer immunotherapy (Muller et al. 2005). Finally, proinflammatory cytokines (e.g., IL-1 and TNF), expressed by infiltrating leukocytes, can activate NF-κB in cancer cells and contribute to their proliferation, survival and metastasis (Balkwill et al. 2001; Balkwill et al. 2005), thus representing potential anticancer targets.

7 Concluding Remarks

Macrophages are key cell components of the inflammatory reactions expressed at the tumor site. Several lines of evidence, ranging from adoptive transfer of cells to genetic manipulations, suggest that myelomonocytic cells can promote tumor invasion and metastasis, although under certain conditions they can express antitumor reactivity.

Several studies have displayed key molecules and pathways driving recruitment and activation of TAM and, more recently, the TAM transcriptome was provided in a murine fibrosarcoma (Biswas et al. 2006). Despite these efforts, TAM functions have been significantly characterized only in animal models, and until now their phenotypic characterization remains only partial in human cancers. Moreover, it is still unknown whether different tumor microenvironments, likely established by different tumor types, may drive different functional phenotypes of TAM and contribute to specific pro- or antitumor activities. As new inflammatory players, recent findings suggest a general pro-tumoral function of toll-like receptors (TLR) in the cancer microenvironment (Mantovani and Garlanda 2006).

Along with TAM recruitment, activation and polarization mechanisms, the functional heterogenicity of TAM should be viewed as an additional level of investigation to develop innovative anticancer strategies.

Acknowledgments. This work was supported by Associazione Italiana Ricerca sul Cancro (AIRC), Italy; by European Community (DC-Thera integrated project, number: LSHB-CT-2004-512074) and by Ministero Istruzione Università Ricerca (MIUR), Italy; Istituto Superiore Sanita' (ISS).

References

Adema, G.J., Hartgers, F., Verstraten, R., de Vries, E., Marland, G., Menon, S., Foster, J., Xu, Y., Nooyen, P., McClanahan, T., Bacon, K.B. and Figdor, C.G. (1997) A dendritic-cell-derived C–C chemokine that preferentially attracts naive T cells. Nature 387, 713–717.

Adini, A., Kornaga, T., Firoozbakht, F. and Benjamin, L.E. (2002) Placental growth factor is a survival factor for tumor endothelial cells and macrophages. Cancer Res. 62, 2749–2752.

Aharinejad, S., Abraham, D., Paulus, P., Abri, H., Hofmann, M., Grossschmidt, K., Schafer, R., Stanley, E.R. and Hofbauer, R. (2002) Colony-stimulating factor-1 antisense treatment suppresses growth of human tumor xenografts in mice. Cancer Res. 62, 5317–5324.

Allavena, P., Sica, A., Vecchi, A., Locati, M., Sozzani, S. and Mantovani, A. (2000) The chemokine receptor switch paradigm and dendritic cell migration: its significance in tumor tissues. Immunol. Rev. 177, 141–149.

Allavena, P., Signorelli, M., Chieppa, M., Erba, E., Bianchi, G., Marchesi, F., Olimpio, C.O., Bonardi, C., Garbi, A., Lissoni, A., de Braud, F., Jimeno, J. and D'Incalci, M. (2005) Anti-inflammatory properties of the novel antitumor agent Yondelis (trabectedin) inhibition of macrophage differentiation and cytokine production. Cancer Res. 65, 2964–2971.

Azenshtein, E., Luboshits, G., Shina, S., Neumark, E., Shahbazian, D., Weil, M., Wigler, N., Keydar, I. and Ben-Baruch, A. (2002) The CC chemokine RANTES in breast carcinoma progression: regulation of expression and potential mechanisms of promalignant activity. Cancer Res. 62, 1093–1102.

Azenshtein, E., Meshel, T., Shina, S., Barak, N., Keydar, I. and Ben-Baruch, A. (2005) The angiogenic factors CXCL8 and VEGF in breast cancer: regulation by an array of pro-malignancy factors. Cancer Lett. 217, 73–86.

Balkwill, F. and Mantovani, A. (2001) Inflammation and cancer: back to Virchow? Lancet 357, 539–545.

Balkwill, F., Charles, K.A. and Mantovani, A. (2005) Smoldering and polarized inflammation in the initiation and promotion of malignant disease. Cancer Cell 7, 211–217.

Balkwill, F. (2004) Cancer and the chemokine network. Nat. Rev. Cancer 4, 540–550.

Bell, D., Chomarat, P., Broyles, D., Netto, G., Harb, G.M., Lebecque, S., Valladeau, J., Davoust, J., Palucka, K.A., and Banchereau, J. (1999) In breast carcinoma tissue, immature dendritic cells reside within the tumor, whereas mature dendritic cells are located in peritumoral areas. J. Exp. Med. 190, 1417–1426.

Bingle, L., Brown, N.J. and Lewis, C.E. (2000) The role of tumour-associated macrophages in tumour progression: implications for new anticancer therapies. J. Pathol. 196, 254–265.

Biswas, S.K., Gangi, L., Paul, S., Schioppa, T., Saccani, A., Sironi, M., Bottazzi, B., Doni, A., Vincenzo, B., Pasqualini, F., Vago, L., Nebuloni, M., Mantovani, A. and Sica, A. (2006) A distinct and unique transcriptional program expressed by tumor-associated macrophages (defective NF-kappaB and enhanced IRF-3/STAT1 activation). Blood 107, 2112–2122.

Borrello, M.G., Alberti, L., Fischer, A., Degl'innocenti, D., Ferrario, C., Gariboldi, M., Marchesi, F., Allavena, P., Greco, A., Collini, P., Pilotti, S., Cassinelli, G., Bressan, P., Fugazzola, L., Mantovani, A. and Pienotti, M.A. (2005) Induction of a proinflammatory program in normal human thyrocytes by the RET/PTC1 oncogene. Proc. Natl. Acad. Sci. U S A. 102, 14825–14830.

Bottazzi, B., Polentarutti, N., Acero, R., Balsari, A., Boraschi, D., Ghezzi, P., Salmona, M. and Mantovani, A. (1983) Regulation of the macrophage content of neoplasms by chemoattractants. Science 220, 210–212.

Bronte, V., Serafini, P., Mazzoni, A., Segal, D.M. and Zanovello, P. (2003) L-arginine metabolism in myeloid cells controls T-lymphocyte functions. Trends Immunol. 24, 302–306.

Bronte, V., Kasic, T., Gri, G., Gallana, K., Borsellino, G., Marigo, I., Battistini, L., Iafrate, M., Prayer-Galetti, T., Pagano, F. and Viola, A. (2005) Boosting antitumor responses of T lymphocytes infiltrating human prostate cancers. J. Exp. Med. 201, 1257–1268.

Carta, L., Pastorino, S., Melillo, G., Bosco, M.C., Massazza, S. and Varesio, L. (2001) Engineering of macrophages to produce IFN-gamma in response to hypoxia. J. Immunol. 166, 5374–5380.

Conti, I. and Rollins, B.J. (2004) CCL2 (monocyte chemoattractant protein-1) and cancer. Semin. Cancer Biol. 14, 149–154.

Coussens, L.M. and Werb, Z. (2002) Inflammation and cancer. Nature 420, 860–867.

Coussens, L.M., Tinkle, C.L., Hanahan, D. and Werb, Z. (2000) MMP-9 supplied by bone marrow-derived cells contributes to skin carcinogenesis. Cell 103, 81–490.

Curiel, T.J., Coukos, G., Zou, L., Alvarez, X., Cheng, P., Mottram, P., Evdemon-Hogan, M., Conejo-Garcia, J.R., Zhang, L., Burow, M., Zhu, Y., Wei, S., Kryczek, I., Daniel, B., Gordon, A., Myers, L., Lackner, A., Disis, M.L., Knutson, K.L., Chen, L. and Zou, W. (2004) Specific recruitment of regulatory T cells in ovarian carcinoma fosters immune privilege and predicts reduced survival. Nat. Med. 10, 942–949.

de Visser, K.E., Eichten, A. and Coussens, L.M. (2006) Paradoxical roles of the immune system during cancer development. Nat. Rev. Cancer 6, 24–37.

Desnues, B., Lepidi, H., Raoult, D. and Mege, J.L. (2005) Whipple disease: intestinal infiltrating cells exhibit a transcriptional pattern of M2/alternatively activated macrophages. J. Infect. Dis. 192, 1642–1646.

Dinapoli, M.R., Calderon, C.L. and Lopez, D.M. (1996) The altered tumoricidal capacity of macrophages isolated from tumor-bearing mice is related to reduced expression of the inducible nitric oxide synthase gene. J. Exp. Med. 183, 1323–1329.

Dinarello, C.A. (2005) Blocking IL-1 in systemic inflammation. J. Exp. Med. 201, 1355–1359.

Dong, Z., Yoneda, J., Kumar, R. and Fidler, I.J. (1998) Angiostatin-mediated suppression of cancer metastases by primary neoplasms engineered to produce granulocyte/macrophage colony-stimulating factor. J. Exp. Med. 188, 755–763.

Duyndam, M.C., Hilhorst, M.C., Schluper, H.M., Verheul, H.M., van Diest, P.J., Kraal, G., Pinedo, H.M. and Boven E. (2002) Vascular endothelial growth factor-165 overexpression stimulates angiogenesis and induces cyst formation and macrophage infiltration in human ovarian cancer xenografts. Am. J. Pathol. 160, 537–548.

Ghassabeh, G.H., De Baetselier, P., Brys, L., Noel, W., Van Ginderachter, J.A., Meerschaut, S., Beschin, A., Brombacher, F. and Raes, G. (2006) Identification of a common gene signature for type II cytokine-associated myeloid cells elicited in vivo in different pathologic conditions. Blood 108, 575–583.

Giraudo, E., Inoue, M. and Hanahan, D. (2004) An amino-bisphosphonate targets MMP-9-expressing macrophages and angiogenesis to impair cervical carcinogenesis. J. Clin. Invest. 114, 623–633.

Goerdt, S. and Orfanos, C.E. (1999) Other functions, other genes: alternative activation of antigen-presenting cells. Immunity 10, 137–142.

Gordon, S. (2003) Alternative activation of macrophages. Nat. Rev. Immunol. 3, 23–35.

Goswami, S., Sahai, E., Wyckoff, J.B., Cammer, M., Cox, D., Pixley, F.J., Stanley, E.R., Segall, J.E. and Condeelis, J.S. (2005) Macrophages promote the invasion of breast carcinoma cells via a colony-stimulating factor-1/epidermal growth factor paracrine loop. Cancer Res. 65, 5278–5283.

Gu, L., Tseng, S., Horner, R.M., Tam, C., Loda, M. and Rollins, B.J. (2000) Control of TH2 polarization by the chemokine monocyte chemoattractant protein-1. Nature 404, 407–411.

Guiducci, C., Vicari, A.P., Sangaletti, S., Trinchieri, G. and Colombo, M.P. (2005) Redirecting in vivo elicited tumor infiltrating macrophages and dendritic cells towards tumor rejection. Cancer Res. 65, 3437–3446.

Hagemann, T., Wilson, J., Burke, F., Kulbe, H., Li, N.F., Pluddemann, A., Charles, K., Gordon, S. and Balkwill, F.R. (2006) Ovarian cancer cells polarize macrophages toward a tumor-associated phenotype. J. Immunol. 176, 5023–5032.

Haghnegahdar, H., Du, J., Wang, D., Strieter, R.M., Burdick, M.D., Nanney, L.B., Cardwell, N., Luan, J., Shattuck-Brandt, R. and Richmond, A. (2000) The tumorigenic and angiogenic effects of MGSA/GRO proteins in melanoma. J. Leukoc. Biol. 67, 53–62.

Hanamoto, H., Nakayama, T., Miyazato, H., Takegawa, S., Hieshima, K., Tatsumi, Y., Kanamaru, A. and Yoshie, O. (2004) Expression of CCL28 by Reed-Sternberg cells defines a major subtype of classical Hodgkin's disease with frequent infiltration of eosinophils and/or plasma cells. Am. J. Pathol. 164, 997–1006.

Hotchkiss, A., Ashton, A.W., Klein, R.S., Lenzi, M.L., Zhu, G.H. and Schwartz E.L. (2003) Mechanisms by which tumor cells and monocytes expressing the angiogenic

factor thymidine phosphorylase mediate human endothelial cell migration. Cancer Res. 63, 527–533.
Jarnicki, A.G., Lysaght, J., Todryk, S. and Mills, K.H. (2006) Suppression of antitumor immunity by IL-10 and TGF-beta-producing T cells infiltrating the growing tumor: influence of tumor environment on the induction of CD4+ and CD8+ regulatory T cells. J. Immunol. 177, 896–904.
Joseph, I.B. and Isaacs, J.T. (1998) Macrophage role in the anti-prostate cancer response to one class of antiangiogenic agents. J. Natl. Cancer. Inst. 90, 1648–1653.
Kleinhans, M., Tun-Kyi, A., Gilliet, M., Kadin, M.E., Dummer, R., Burg, G. and Nestle, F.O. (2003) Functional expression of the eotaxin receptor CCR3 in CD30+ cutaneous T-cell lymphoma. Blood 101, 1487–1493.
Klimp, A.H., Hollema, H., Kempinga, C., van der Zee, A.G., de Vries, E.G. and Daemen, T. (2001) Expression of cyclooxygenase-2 and inducible nitric oxide synthase in human ovarian tumors and tumor-associated macrophages. Cancer Res. 61, 7305–7309.
Kortylewski, M., Kujawski, M., Wang, T., Wei, S., Zhang, S., Pilon-Thomas, S., Niu, G., Kay, H., Mule, J., Kerr, W.G., Jove, R., Pardoll, D. and Yu, H. (2005) Inhibiting Stat3 signaling in the hematopoietic system elicits multicomponent antitumor immunity. Nat. Med. 11, 1314–1321.
Kouno, J., Nagai, H., Nagahata, T., Onda, M., Yamaguchi, H., Adachi, K., Takahashi, H., Teramoto, A. and Emi, M. (2004) Up-regulation of CC chemokine, CCL3L1, and receptors, CCR3, CCR5 in human glioblastoma that promotes cell growth. J. Neurooncol. 70, 301–307.
Kusmartsev, S. and Gabrilovich, D.I. (2005) STAT1 signaling regulates tumor-associated macrophage-mediated T cell deletion. J. Immunol. 174, 4880–4891.
Li, A., Varney. M.L. and Singh, R.K. (2004) Constitutive expression of growth regulated oncogene (gro) in human colon carcinoma cells with different metastatic potential and its role in regulating their metastatic phenotype. Clin. Exp. Metastasis 21, 571–579.
Lin, E.Y., Nguyen, A.V., Russell, R.G. and Pollard, J.W. (2001) Colony-stimulating factor 1 promotes progression of mammary tumors to malignancy. J. Exp. Med. 193, 727–740.
Locati, M., Deuschle, U., Massardi, M.L., Martinez, F.O., Sironi, M., Sozzani, S., Bartfai, T. and Mantovani, A. (2002) Analysis of the gene expression profile activated by the CC chemokine ligand 5/RANTES and by lipopolysaccharide in human monocytes. J. Immunol. 168, 3557–3562.
Mantovani, A., Sozzani, S., Locati, M., Allavena, P. and Sica, A. (2002) Macrophage polarization: tumor-associated macrophages as a paradigm for polarized M2 mononuclear phagocytes. Trends Immunol. 23, 549–555.
Mantovani, A., Sica, A., Sozzani, S., Allavena, P., Vecchi, A. and Locati, M. (2004) The chemokine system in diverse forms of macrophage activation and polarization. Trends Immunol. 25, 677–686.
Mantovani, A. (1999) The chemokine system: redundancy for robust outputs. Immunol. Today 20, 254–257.
Mantovani, A., Bottazzi, B., Colotta, F., Sozzani, S. and Ruco, L. (1992) The origin and function of tumor-associated macrophages. Immunol. Today 13, 265–270.
Mantovani, A and Garlanda, C. (2006) Inflammation and multiple myeloma: the Toll connection. Leukemia 20, 937–938.
Matsushima, K., Larsen, C.G., DuBois, G.C. and Oppenheim, J.J. (1999) Purification and characterization of a novel monocyte chemotactic and activating factor produced by a human myelomonocytic cell line. J. Exp. Med. 169, 1485–1490.
Monti, P., Leone, B.E., Marchesi, F., Balzano, G., Zerbi, A., Scaltrini, F., Pasquali, C., Calori, G., Pessi, F., Sperti, C., Di Carlo, V., Allavena, P. and Piemonti, L. (2003)

The CC chemokine MCP-1/CCL2 in pancreatic cancer progression: regulation of expression and potential mechanisms of antimalignant activity. Cancer Res. 63, 7451–7461.

Mori, K., Chano, T., Yamamoto, K., Matsusue, Y. and Okabe, H. (2004) Expression of macrophage inflammatory protein-1alpha in Schwann cell tumors. Neuropathology 24, 131–135.

Muller, A.J., DuHadaway, J.B., Donover, P.S., Sutanto-Ward, E. and Prendergast, G.C. (2005) Inhibition of indoleamine 2,3-dioxygenase, an immunoregulatory target of the cancer suppression gene Bin1, potentiates cancer chemotherapy. Nat. Med. 11, 312–319.

Nesbit, M., Schaider, H., Miller, T.H. and Herlyn, M. (2001) Low-level monocyte chemoattractant protein-1 stimulation of monocytes leads to tumor formation in nontumorigenic melanoma cells. J. Immunol. 166, 6483–6490.

Noel, W., Raes, G., Hassanzadeh Ghassabeh, G., De Baetselier, P. and Beschin, A. (2004) Alternatively activated macrophages during parasite infections. Trends Parasitol. 20, 126–133.

Nowicki, A., Szenajch, J., Ostrowska, G., Wojtowicz, A., Wojtowicz, K., Kruszewski, A.A., Maruszynski, M., Aukerman, S.L. and Wiktor-Jedrzejczak, W. (1996) Impaired tumor growth in colony-stimulating factor 1 (CSF-1)-deficient, macrophage-deficient op/op mouse: evidence for a role of CSF-1-dependent macrophages in formation of tumor stroma. Int. J. Cancer 65, 112–119.

Paik, S., Shak, S., Tang, G., Kim, C., Baker, J., Cronin, M., Baehner, F.L., Walker, M.G., Watson, D., Park, T., Hiller, W., Fisher, E.R., Wickerham, D.L., Bryant, J. and Wolmark, N. (2004) A multigene assay to predict recurrence of tamoxifen-treated, node-negative breast cancer. N. Engl. J. Med. 351, 2817–2826.

Payne, A.S. and Cornelius, L.A. (2002) The role of chemokines in melanoma tumor growth and metastasis. J. Invest. Dermatol. 118, 915–922.

Phillips, R.J., Mestas, J., Gharaee-Kermani, M., Burdick, M.D., Sica, A., Belperio, J.A., Keane, M.P. and Strieter, R.M. (2005) Epidermal growth factor and hypoxia-induced expression of CXC chemokine receptor 4 on nonsmall cell lung cancer cells is regulated by the phosphatidylinositol 3-kinase/PTEN/AKT/mammalian target of rapamycin signaling pathway and activation of hypoxia inducible factor-1alpha. J. Biol. Chem. 280, 22473–22481.

Pollard, J.W. (2004) Tumour-educated macrophages promote tumour progression and metastasis Nat. Rev. Cancer 4, 71–78.

Rauh, M.J., Sly, L.M., Kalesnikoff, J., Hughes, M.R., Cao, L.P., Lam, V. and Krystal, G. (2004) The role of SHIP1 in macrophage programming and activation. Biochem. Soc. Trans. 32, 785–788.

Rollins, B. (1999) *Chemokines and Cancer*. Humana Press, Totowa, NJ.

Ruckes, T., Saul, D., Van Snick, J., Hermine, O. and Grassmann, R. (2001) Autocrine antiapoptotic stimulation of cultured adult T-cell leukemia cells by overexpression of the chemokine I-309. Blood 98, 1150–1159.

Saji, H., Koike, M., Yamori, T., Saji, S., Seiki, M., Matsushima, K. and Toi, M. (2001) Significant correlation of monocyte chemoattractant protein-1 expression with neovascularization and progression of breast carcinoma. Cancer 92, 1085–1091.

Sakaguchi, S. (2005) Naturally arising Foxp3-expressing CD25+CD4+ regulatory T cells in immunological tolerance to self and non-self. Nat Immunol. 6, 345–352.

Sangaletti, S., Stoppacciaro, A., Guiducci, C., Torrisi, M.R. and Colombo, M.P. (2003) Leukocyte, rather than tumor-produced SPARC, determines stroma and collagen type IV deposition in mammary carcinoma. J. Exp. Med. 198, 1475–1485.

Scala, S., Ottaiano, A., Ascierto, P.A., Cavalli, M., Simeone, E., Giuliano, P., Napolitano, M., Franco, R., Botti, G. and Castello, G. (2005) Expression of CXCR4 predicts poor prognosis in patients with malignant melanoma. Clin. Cancer Res. 11, 1835–1841.

Scarpino, S., Stoppacciaro, A., Ballerini, F., Marchesi, M., Prat, M., Stella, M.C., Sozzani, S., Allavena, P., Mantovani, A. and Ruco, L.P. (2000) Papillary carcinoma of the thyroid: hepatocyte growth factor (HGF) stimulates tumor cells to release chemokines active in recruiting dendritic cells. Am. J. Pathol. 156, 831–837.

Schoppmann, S.F., Birner, P., Stockl, J., Kalt, R., Ullrich, R., Caucig, C., Kriehuber, E., Nagy, K., Alitalo, K. and Kerjaschki, D. (2002) Tumor-associated macrophages express lymphatic endothelial growth factors and are related to peritumoral lymphoangiogenesis. Am. J. Pathol. 161, 947–956.

Schutyser, E., Struyf, S., Proost, P., Opdenakker, G., Laureys, G., Verhasselt, B., Peperstraete, L., Van de Putte, I., Saccani, A., Allavena, P., Mantovani, A. and Van Damme, J. (2002) Identification of biologically active chemokine isoforms from ascitic fluid and elevated levels of CCL18/pulmonary and activation-regulated chemokine in ovarian carcinoma. J. Biol. Chem. 277, 24584–24593.

Scotton, C.J., Martinez, F.O., Smelt, M.J., Sironi, M., Locati, M., Mantovani, A. and Sozzani, S. (2005) Transcriptional profiling reveals complex regulation of the monocyte IL-1 beta system by IL-13. J. Immunol. 174, 834–845.

Sessa, C., De Braud, F., Perotti, A., Bauer, J., Curigliano, G., Noberasco, C., Zanaboni, F., Gianni, L., Marsoni, S., Jimeno, J., D'Incalci, M., Dall'o, E. and Colombo, N. (2005) Trabectedin for women with ovarian carcinoma after treatment with platinum and taxanes fails. J. Clin. Oncol. 23, 1867–1874.

Sica, A., Saccani, A., Bottazzi, B., Polentarutti, N., Vecchi, A., van Damme, J. and Mantovani, A. (2000) Autocrine production of IL-10 mediates defective IL-12 production and NF-kappa B activation in tumor-associated macrophages. J. Immunol. 164, 762–767.

Sinha, P., Clements, V.K. and Ostrand-Rosenberg, S. (2005) Reduction of myeloid-derived suppressor cells and induction of M1 macrophages facilitate the rejection of established metastatic disease. J. Immunol. 174, 636–645.

Smith, J.R., Braziel, R.M., Paoletti, S., Lipp, M., Uguccioni, M. and Rosenbaum, J.T. (2003) Expression of B-cell-attracting chemokine 1 (CXCL13) by malignant lymphocytes and vascular endothelium in primary central nervous system lymphoma. Blood 101, 815–821.

Takahashi, H., Tsuda, Y., Takeuchi, D., Kobayashi, M., Herndon, D.N. and Suzuki, F. (2004) Influence of systemic inflammatory response syndrome on host resistance against bacterial infections. Crit. Care Med. 32, 1879–1885.

Teichmann, M., Meyer, B., Beck, A. and Niedobitek, G. (2005) Expression of the interferon-inducible chemokine IP-10 (CXCL10), a chemokine with proposed anti-neoplastic functions, in Hodgkin lymphoma and nasopharyngeal carcinoma. J. Pathol. 20, 68–75.

Teruya-Feldstein, J., Tosato, G. and Jaffe, E.S. (2000) The role of chemokines in Hodgkin's disease. Leuk. Lymphoma 38, 363–371.

Torroella-Kouri, M., Ma, X., Perry, G., Ivanova, M., Cejas, P.J., Owen, J.L., Iragavarapu-Charyulu, V. and Lopez, D.M. (2005) Diminished expression of transcription factors nuclear factor kappaB and CCAAT/enhancer binding protein underlies a novel tumor evasion mechanism affecting macrophages of mammary tumor-bearing mice. Cancer Res. 65, 10578–10584.

Tsuda, Y., Takahashi, H., Kobayashi, M., Hanafusa, T., Herndon, D.N. and Suzuki, F. (2004) Three different neutrophil subsets exhibited in mice with different susceptibilities to infection by methicillin-resistant Staphylococcus aureus. Immunity 21, 215–226.

Ueno, T., Toi, M., Saji, H., Muta, M., Bando, H., Kuroi, K., Koike, M., Inadera, H. and Matsushima, K. (2000) Significance of macrophage chemoattractant protein-1 in macrophage recruitment, angiogenesis, and survival in human breast cancer. Clin. Cancer Res. 6, 3282–3389.

Uyttenhove, C., Pilotte, L., Theate, I., Stroobant, V., Colau, D., Parmentier, N., Boon, T. and Van den Eynde, B.J. (2003) Evidence for a tumoral immune resistance mechanism based on tryptophan degradation by indoleamine 2,3-dioxygenase. Nat. Med. 9, 1269–1274.

Van Damme, J., Proost, P., Lenaerts, J.P. and Opdenakker, G. (1992) Structural and functional identification of two human, tumor-derived monocyte chemotactic proteins (MCP-2 and MCP-3) belonging to the chemokine family. J. Exp. Med. 176, 59–65.

Van den Brule, F., Califice, S., Garnier, F., Fernandez, P.L., Berchuck, A. and Castronovo, V. (2003) Galectin-1 accumulation in the ovary carcinoma peritumoral stroma is induced by ovary carcinoma cells and affects both cancer cell proliferation and adhesion to laminin-1 and fibronectin. Lab. Invest. 83, 377–386.

Vermeer, M.H., Dukers, D.F., ten Berge, R.L., Bloemena, E., Wu, L., Vos, W., de Vries, E., Tensen, C.P., Meijer, C.J. and Willemze, R. (2002) Differential expression of thymus and activation regulated chemokine and its receptor CCR4 in nodal and cutaneous anaplastic large-cell lymphomas and Hodgkin's disease. Mod. Pathol. 15, 838–844.

Vicari, A.P. and Caux, C. (2002) Chemokines in cancer. Cytokine Growth Factor Rev. 13, 143–154.

Wang, T., Niu, G., Kortylewski, M., Burdelya, L., Shain, K., Zhang, S., Bhattacharya, R., Gabrilovich, D., Heller, R., Coppola, D., Dalton, W., Jove, R., Pardoll, D. and Yu, H. (2004) Regulation of the innate and adaptive immune responses by Stat-3 signaling in tumor cells. Nat. Med. 10, 48–54.

Wynn, T.A. (2004) Fibrotic disease and the T(H)1/T(H)2 paradigm. Nat. Rev. Immunol. 4, 583–594.

Zhu, Z., Zheng, T., Homer, R.J., Kim, Y.K., Chen, N.Y., Cohn, L., Hamid, Q. and Elias, J.A. (2004) Acidic mammalian chitinase in asthmatic Th2 inflammation and IL-13 pathway activation. Science 304, 1678–1682.

Yang, J. and Richmond, A. (2001) Constitutive IkappaB kinase activity correlates with nuclear factor-kappaB activation in human melanoma cells. Cancer Res. 61, 4901–4909.

Yi, F., Jaffe, R. and Prochownik, E.V. (2003) The CCL6 chemokine is differentially regulated by c-Myc and L-Myc, and promotes tumorigenesis and metastasis. Cancer Res. 63, 2923–2932.

Yoshimura, T., Robinson, E.A., Tanaka, S., Appella, E., Kuratsu, J. and Leonard, E.J. (1989) Purification and aminoacid analysis of two human glioma-derived monocyte chemoattractants. J. Exp. Med. 169, 1449–1459.

Chapter 6
The AP-2α Transcription Factor Regulates Tumor Cell Migration and Apoptosis

Francesca Orso, Michela Fassetta, Elisa Penna, Alessandra Solero, Katia De Filippo, Piero Sismondi, Michele De Bortoli and Daniela Taverna*

Institute for Cancer Research and Treatment (IRCC), Department of Oncology. Science, University of Torino, Str. Prov. 142 – Km 3.95, 10060 Candiolo (To), Italy,
*daniela.taverna@unito.it

Abstract. AP-2 proteins are a family of developmentally-regulated transcription factors. They are encoded by five different genes (α, β, γ, δ, and ε) but they share a common structure. *AP-2* plays relevant roles in growth, differentiation, and adhesion by controlling the transcription of specific genes. Evidence shows that the *AP-2* genes are involved in tumorigenesis and for instance, they act as tumor suppressors in melanomas and mammary carcinomas. Here we investigated the function of the AP-2α protein in cancer formation and progression focusing on apoptosis and migration. We introduced AP-2α-specific siRNA (as oligos or in retroviruses) in HeLa or MCF-7 human tumor cells and obtained a pronounced down-modulation of AP-2α mRNA and protein levels. In these cells, we observed a significant reduction of chemotherapy-induced apoptosis, migration, and motility and an increase in adhesion suggesting a major role of AP-2α during cancer treatment and progression (migration and invasion). We have data suggesting that migration is, at least in part, regulated by secreted factors. By performing a whole genome microarray analysis of the tumor cells expressing AP-2α siRNA, we identified several AP-2α -regulated genes involved in apoptosis and migration such as FAST kinase, osteopontin, caspase 9, members of the TNF family, laminin alpha 1, collagen type XII, alpha 1, and adam.

1 Introduction

The AP-2 family of transcription factors consists of five closely related proteins of Mr 50,000, AP-2α, β, γ, δ, and ε (Feng and Williams, 2003; Hilger-Eversheim et al., 2000; Zhao et al., 2001) encoded by distinct genes. These transcription factors can form homoor heterodimers via helix–span–helix motifs and transactivate their target genes by binding to GC-rich consensus sequences in the promoter regions (Williams and Tjian, 1991). AP-2 factors orchestrate a variety of cell processes including apoptosis, cell growth, cell adhesion, tissue differentiation and tumorigenesis

(Hilger-Eversheim et al., 2000). Generation and analysis of mice deficient in AP-2 α, β or γ have indicated important roles for AP-2 factors in the development of neural crest, urogenital, and epidermal tissues during embryogenesis (Moser et al., 1995; Moser et al., 1997; Schorle et al., 1996; Werling and Schorle, 2002). The crucial role of *AP-2* genes in regulating gene expression is highlighted by the embryonic lethality of all these genetically modified mice. Some of the genes regulated by AP-2 transcription factors are p21WAF/CIP (Zeng et al., 1997), transforming growth factor-α (Wang et al., 1997), estrogen receptor α (McPherson et al., 1997), keratinocyte specific genes (Leask et al., 1991), c-KIT (Huang et al., 1998), adhesion molecules such as MCAM/MUC18 or integrins (Suyama et al., 2002), type IV collagenase/gelatinase/MMP-2, E-cadherin, VEGF (Nyormoi and Bar-Eli, 2003) and ERBB-2 (reviewed in Hilger-Eversheim et al., 2000). Several studies suggest that AP-2 plays a role in tumorigenesis acting as oncogene or tumor suppressor depending on the type of tumor (Hilger-Eversheim et al., 2000). Here AP-2 regulates gene expression directly by binding to the regulatory regions of some of the genes listed above or indirectly displaying functional protein–protein interactions with other transcription factors such as c-myc, pRB, and p53 (Hilger-Eversheim et al., 2000).

2 Results

2.1 *Downregulation of AP-2α Expression in Tumor Epithelial Cells*

Many tumor cells and human epithelial tumors show downregulation of the AP-2transcription factors (Hilger-Eversheim et al., 2000). Here we planned to downmodulate AP-2α in tumor cells by using specific siRNA and study tumor formation and progression. HeLa (cervix adenocarcinoma) or MCF7 (breast adenocarcinoma) human cells were first transiently transfected either with generic nonsilencing (NS, Qiagen) or specific AP-2α siRNA oligos (Qiagen, sequences in Figure 1) and analyzed for protein expression 48 h after transfection by western blot (Figure 2a). Alternatively, the cells were transduced with retroviruses and selected to obtain stable expression of pSUPERretro.puro-OLIGO2 or pSUPER retro.puro-OLIGO4. The empty pSUPERretro.puro was used as control. Protein expression was also analyzed by western blot (Figure 2b,c). A very significant reduction of AP-2α expression was observed when the cells were either transfected with OLIGO2 or OLIGO4 siRNA oligos or transduced with pSUPERretro.puro-OLIGO2 or pSUPER retro.puro-OLIGO4 expression vectors compared with control cells (Figure 2).

6. The AP-2α Transcription Factor Regulates Tumor Cell Migration

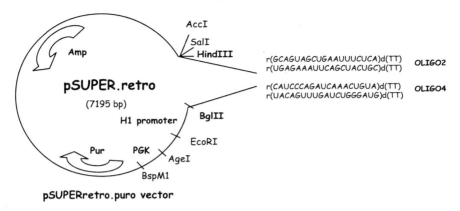

FIGURE 1. **Constructs.** The indicated OLIGO2 and OLIGO4 siRNA oligonucleotide sequences targeting the human AP-2α mRNA were cloned in the pSUPERretro.puro expression vector (OligoEngine Inc.) using the *Hin*dIII and *Bgl*II cloning sites.

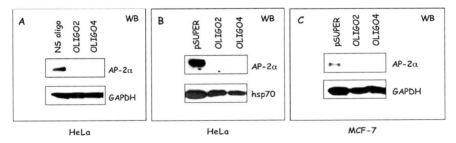

FIGURE 2. **Downregulation of AP-2α protein expression by specific siRNA.** HeLa cells were transiently transfected (a) either with generic nonsilencing (NS, Qiagen) or specific (OLIGO2 or OLIGO4, Qiagen) AP-2α siRNA oligos and the AP-2α protein expression was analyzed by western blot. In other cases HeLa (b) or MCF7 (c) cells were infected with retroviruses containing pSUPERretro.puro-OLIGO2 or pSUPERretro.puro-OLIGO4 expression vector to obtain stable expression of OLIGO2 or OLIGO4 AP-2α siRNA and analyzed for protein expression. Cells infected with the empty pSUPERretro.puro vector (pSUPER) were used as controls in (b) and (c). The expression of glyceraldheyde-3-phosphate dehydrogenase (GAPDH) or heat shock protein 70 (hsp70) was used to control for protein loading. The OLIGO2 and OLIGO4 sequences induced a pronounced downmodulation of AP-2α in both cell lines used.

2.2 *AP-2α Downmodulation Leads to Reduced Apoptosis*

Transduced HeLa or MCF7 cells expressing either an empty pSUPER retro.puro (pSUPER) or a pSUPERretro.puro-OLIGO2 (OLIGO2) vector were left untreated (CTR) or stimulated with staurosporine (STS) (Figure 3

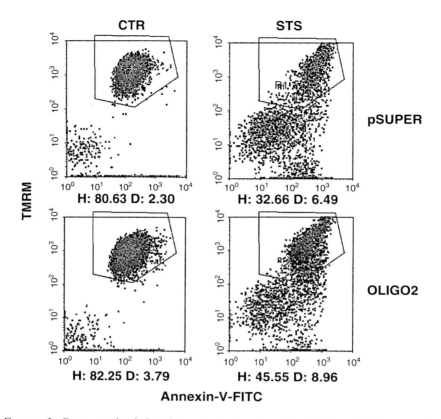

FIGURE 3. **Staurosporine-induced apoptosis is reduced in AP-2α siRNA-expressing cells.** Retrovirus infected HeLa cells expressing either an empty pSUPERretro.puro (pSUPER) or a pSUPERretro.puro-OLIGO2 (OLIGO2) vector were left untreated (CTR) or stimulated with Staurosporine (STS) for 24 h to induce apoptosis. FACS analysis (Annexin V/TMRM/Propidium Iodide) shows reduced cell death in cells expressing low levels of AP-2α (OLIGO2) compared with control cells (pSUPER) following STS treatment. Cells displaying mitochondrial depolarization are in the lower part of each plot (reduced TMRM) whereas cells presenting phosphatidylserine on their surface are in the right quarter of the diagrams (increased Annexin V-FITC stainings). We report the percentage of healthy cells (H, delimited by the gate) and of propidium iodide-permeable dead cells (D, not shown in the diagrams). One representative experiment is shown.

and Table 1) or paclitaxel (PTX) (Table 1) for 24 or 48 or 72 h to induce apoptosis. FACS analysis was then performed and the percentage of healthy (H) or dead (D) cells was calculated. Reduced cell death was observed in cells with pronounced downregulation of AP-2α protein

TABLE 1. Analysis of cell death.

Cells	Stimulus	Conc.	Time	% Of apoptotic cells pSUPER	OLIGO2	n
HeLa	STS	25 ng/ml	24h	41,9	29,8	6
	PTX	20 nM	72h	62,4	37,1	2
MCF-7	STS	25 ng/ml	24h	48,3	33,9	2
	PTX	20 nM	48h	22,5	9,6	3

HeLa or MCF-7 cells transduced with retroviruses carrying either the empty pSUPERretro.puro (pSUPER) or a pSUPERretro.puro-OLIGO2 (OLIGO2) vector were stimulated either with staurosporine (STS) or with paclitaxel (PTX) at the indicated concentration and time. The percentage of apoptotic cells measured by FACS analysis is indicated. (n) is the number of experiments performed.

expression (OLIGO2) compared with control cells (pSUPER) (Figure 3 and Table 1).

Microarray analysis was performed on HeLa cells transiently transfected with OLIGO2 or NS siRNA oligos and several apoptosis related genes turned out to be regulated by AP-2α (see Table 2). Validation of microarray results by Real Time PCR of some individual genes is shown in Table 2.

2.3 Migration and Motility are Reduced in Cells with Low AP-2α Expression While Adhesion is Increased

Retrovirus infected HeLa cells expressing either an empty pSUPER retro.puro or a pSUPERretro.puro-OLIGO2 or pSUPER retro.puro-OLIGO4 expression vector were used to analyze migration (Figure 4a), motility (Figure 4b), or adhesion (Figure 4c). Reduced migration or motility or increased adhesion over collagen or fibronectin were observed in pSUPERretro.puro-OLIGO2-expressing cells compared with cells containing the empty vector (pSUPER). To test whether the decreased migration was cell autonomous or due to secreted factors regulated by AP-2α, we analyzed the effect of serum free-conditioned medium (CM) produced either by pSUPER- or OLIGO2- or OLIGO4-expressing cells on control pSUPER cells in a transwell assay as described in Figure 4d. Reduced migration of pSUPER cells was observed only when CM produced from OLIGO2 or OLIGO4-expressing cells was used but not when control CM (produced by pSUPER cells) was added suggesting that decreased migration is at least in part due to secreted molecules. From the microarray analysis we performed on HeLa cells (see above) several migration and adhesion related genes turned out to be regulated by AP-2α (see Table 2). The results of the validation by real time PCR are also shown in Table 2.

TABLE 2. AP-2α modulated genes: a microarray analysis.

Function	Uniqid	Gene name	Array FC	RT FC
Apoptosis	NM_033015	FAST kinase	−2.2	−2,5
	NM_001165	baculoviral IAP repeat-containing 3	−2,1	−2,5
	NM_000700	annexin a1	1.7	
	NM_000389	cyclin-dependent kinase inhibitor 1A (p21, Cip1)	−2.1	−4.3
	NM_153012	tumor necrosis factor superfamily, member 12	−1.6	
	NM_002546	tumor necrosis factor superfamily, member 11b	1.7	
	NM_000582	secreted phosphoprotein 1(osteopontin)	1.7	
	NM_001229	caspase 9, apoptosis-related cysteine peptidase	−1.5	
	NM_003806	harakiri, bcl2 interacting protein	1.8	
	NM_001252	tumor necrosis factor superfamily member 7	−1.9	
Adhesion/ Migration	NM_003881	WNT1 inducible signaling pathway protein 2	−2	−2.5
	NM_005397	podocalyxin-like	2	
	U83115	absent in melanoma 1	2.4	
	NM_080927	neuropilin-like protein (DCBLD2)	5.8	3.2
	NM_000358	transforming growth factor, beta-induced, 68kD	−1.6	
	BC042028	integrin, beta 8	1.7	
	NM_000582	secreted phosphoprotein 1 (ostopontin)	1.7	
	NM_005559	laminin, alpha 1	1.5	
	NM_004370	collagen, type XII, alpha 1	−1.7	
	NM_004598	sparc/osteonectin, cwcv and kazal-like domains	−1.5	
	NM_006988	adam metallopeptidase with thrombospondin type 1 motif 1	1.7	

HeLa cells were transiently transfected with either AP-2α (OLIGO2) or nonsilencing siRNA (NS) siRNA oligos. mRNA was prepared 48 h later and used to perform a whole genome microarray analysis of gene expression (Human Agilent 44 K). Here we compared gene expression of OLIGO2-containing cells versus NS-containing cells and differential gene expression was found. Some AP-2α modulated gene are shown here. Cut-off used: p-value <0.01; Array Fold Change (FC) over 1.5. Some microarray results were validated by Real Time (RT) PCR and the relative FC is shown (RT FC).

2.4 AP-2α Binds to Promoters of Genes Identified by Microarray Analysis

In order to verify the binding of AP-2α to the regulatory regions of some genes identified by microarray analysis (Agilent) we performed chromatin immunoprecipitation (ChIP) assays in 293T cells. Here we tested and confirmed the direct binding of AP-2α to the p21 and DCBLD2 (neuropilin-like protein)

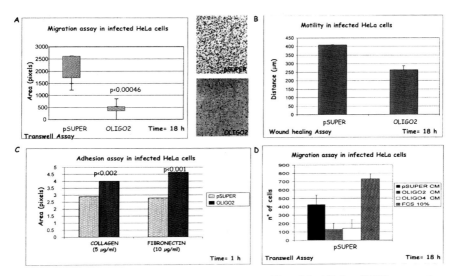

FIGURE 4. **Migration, motility, and adhesion are affected in AP-2α siRNA-expressing cells.** Retrovirus infected HeLa cells expressing either an empty pSUPERretro.puro or a pSUPERretro.puroOLIGO2 or pSUPERretro.puro-OLIGO4 expression vector were used in a transwell assay to analyze migration (**a**) at 18 h while motility (**b**) was tested by wound healing assay at 18 h. In both cases a reduction of migration or motility was observed in pSUPERretro.puro-OLIGO2-expressing cells compared with cells containing the empty vector (pSUPER). Adhesion over collagen or fibronectin was measured 1 h after plating the cells and an increase in adhesion was observed for both matrices when the cells expressed only a low level of AP-2α (OLIGO2) compared with control cells (pSUPER) (**c**). To test whether migration was due to secreted factors regulated by AP-2α we analyzed migration of pSUPER-containing HeLa cells by transwell assays putting serum free-conditioned medium (CM) in the lower chambers (**d**). Here CM was derived from overnight culture of either pSUPER- or OLIGO2- or OLIGO4-expressing cells. Migration over regular medium containing 10% FCS was used as control. Each experiment was performed in triplicate and repeated twice. Representative experiments are shown here. The decreased cell migration or motility and the increased adhesion on collagen or fibronectin were statistically significant as measured by a two tails student t test ($p < 0.05$).

promoter regions (Figure 5) containing the putative AP-2 sites identified by the RRE program (http://www.bioinformatica.unito.it/bioinformatics/rre/rre.html). 3B5 or C-18 antibodies were used to immunoprecipitate AP-2α-bound chromatin fragments while IgG were used instead of an AP-2α antibody, as negative control. p21 and DCBLD2 regulatory regions in the immunoprecipitate were detected by PCR. RNA pol II represented the positive ChIP control used while the no-template PCR reaction represented our negative PCR control.

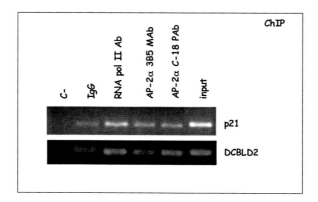

FIGURE 5. **Binding of AP-2α to p21 and DCBLD2 promoters.** Binding of AP-2α to the p21 and DCBLD2 (neuropilin-like protein) regulatory regions was analyzed in 293T cells by chromatin immuno precipitation (ChIP) assay. Chromatin was isolated and immunoprecipitated using two different AP-2α antibodies (3B5 and C-18, Santa Cruz Biotech) or IgG as negative control. DNA purified from the immunoprecipitated fractions was subjected to PCR amplification for the regulatory regions of the *p21* and *DCBLD2* genes containing the putative AP-2 sites using specific primers. ChIP was performed following the ChIP-IT™ Enzymatic kit instructions (Activemotif). anti-RNA *pol*II antibody was used as positive control; input: control of DNA amount. C-represents the no-template, negative control for the PCR.

3 Conclusions

From our results we can conclude that by using AP-2α siRNA we downmodulated AP-2α expression in epithelial tumor cells. The downregulation of AP-2α led to reduced migration and motility and to increased adhesion over collagen and fibronectin. We observed that migration was at least in part modulated by secreted factors. Apoptosis induced by staurosporine or taxol was also reduced. These results suggest that AP-2α plays a relevant role in tumor progression and during cancer treatment in chemotherapy. It is now essential to dissect the molecular mechanisms involved in those biological processes. For that purpose we performed microarray analysis to identify genes modulated by AP-2α in a direct or indirect manner and several genes involved in migration and apoptosis were identified. Direct binding of AP-2α to the regulatory regions of two genes identified by microarray was tested and observed by ChIP analysis.

Acknowledgments. This work was supported by grants from the Regione Piemonte (CIPE2004), University of Torino (Local Research Founding 2004, 2005) and the Italian MURST. FO and MF

References

Feng, W., and Williams, T. (2003). Cloning and characterization of the mouse *AP-2 epsilon* gene: a novel family member expressed in the developing olfactory bulb. Mol Cell Neurosci *24*, 460–475.

Hilger-Eversheim, K., Moser, M., Schorle, H., and Buettner, R. (2000). Regulatory roles of AP-2 transcription factors in vertebrate development, apoptosis and cell-cycle control. Gene *260*, 1–12.

Huang, S., Jean, D., Luca, M., Tainsky, M. A., and Bar-Eli, M. (1998). Loss of AP-2 results in downregulation of c-KIT and enhancement of melanoma tumorigenicity and metastasis. Embo J *17*, 4358–4369.

Leask, A., Byrne, C., and Fuchs, E. (1991). Transcription factor AP2 and its role in epidermal-specific gene expression. Proc Natl Acad Sci U S A *88*, 7948–7952.

McPherson, L. A., Baichwal, V. R., and Weigel, R. J. (1997). Identification of ERF-1 as a member of the AP2 transcription factor family. Proc Natl Acad Sci U S A *94*, 4342–4347.

Moser, M., Imhof, A., Pscherer, A., Bauer, R., Amselgruber, W., Sinowatz, F., Hofstadter, F., Schule, R., and Buettner, R. (1995). Cloning and characterization of a second AP-2 transcription factor: AP-2 beta. Development *121*, 2779–2788.

Moser, M., Pscherer, A., Roth, C., Becker, J., Mucher, G., Zerres, K., Dixkens, C., Weis, J., Guay-Woodford, L., Buettner, R., and Fassler, R. (1997). Enhanced apoptotic cell death of renal epithelial cells in mice lacking transcription factor AP-2beta. Genes Dev *11*, 1938–1948.

Nyormoi, O., and Bar-Eli, M. (2003). Transcriptional regulation of metastasis-related genes in human melanoma. Clin Exp Metastasis *20*, 251–263.

Schorle, H., Meier, P., Buchert, M., Jaenisch, R., and Mitchell, P. J. (1996). Transcription factor AP-2 essential for cranial closure and craniofacial development. Nature *381*, 235–238.

Suyama, E., Minoshima, H., Kawasaki, H., and Taira, K. (2002). Identification of AP-2-regulated genes by macroarray profiling of gene expression in human A375P melanoma. Nucleic Acids Res Suppl, 247–248.

Wang, X., Yang, Y., and Adamo, M. L. (1997). Characterization of the rat insulin-like growth factor I gene promoters and identification of a minimal exon 2 promoter. Endocrinology *138*, 1528–1536.

Werling, U., and Schorle, H. (2002). Transcription factor gene AP-2 gamma essential for early murine development. Mol Cell Biol *22*, 3149–3156.

Williams, T., and Tjian, R. (1991). Characterization of a dimerization motif in AP-2 and its function in heterologous DNA-binding proteins. Science *251*, 1067–1071.

Zeng, G., Dave, J. R., and Chiang, P. K. (1997). Induction of proto-oncogenes during 3-deazaadenosine-stimulated differentiation of 3T3-L1 fibroblasts to adipocytes: mimicry of insulin action. Oncol Res *9*, 205–211.

Zhao, F., Satoda, M., Licht, J. D., Hayashizaki, Y., and Gelb, B. D. (2001). Cloning and characterization of a novel mouse AP-2 transcription factor, AP-2delta, with unique DNA binding and transactivation properties. J Biol Chem *276*, 40755–40760.

Chapter 7
Modulatory Actions of Neuropeptide Y on Prostate Cancer Growth: Role of MAP Kinase/ERK $^1/_2$ Activation

Massimiliano Ruscica, Elena Dozio, Marcella Motta, and Paolo Magni*

Center for Endocrinological Oncology, Department of Endocrinology, University of Milan, Milan, Italy,
*paolo.magni@unimi.it

Abstract. Neuroendocrine molecules play a significant role in the progression of human prostate cancer (PCa) and its neuroendocrine differentiation has been associated to a worse prognosis. Evidence exists that, among these molecules, the pleiotropic neuropeptide Y (NPY) and the related receptors may play a role in the normal prostate as well as in the progression of human PCa, which represents one of the most common malignant diseases among men in the Western world. The role of NPY in PCa biology appears to vary in different in vitro human PCa cell systems, since it has been found to reduce the proliferation of LNCaP and DU145 cells, but to stimulate the growth of PC3 cells. These effects are mediated mainly by the NPY Y1 receptor and are associated with a clone-specific pattern of intracellular signaling activation, including a peculiar time-course of MAPK/ERK1/2 phosphorylation (long-lasting in DU145 and transient in PC3 cells). In conclusion, several studies support the concept that NPY and the related receptors are overexpressed in PCa and may play a relevant role in PCa progression. The diagnostic and therapeutical value of targeting the NPY system in PCa will be evaluated in future studies.

1 Introduction

Neuropeptide Y (NPY) is a highly conserved 36-aminoacid residue peptide, belonging to the pancreatic polypeptide family, which has been involved in the regulation of a wide range of physiological functions, including energy metabolism and food intake, reproduction, neuroendocrine mechanisms, and cardiovascular and cognitive processes (Magni 2003; Pedrazzini et al. 2003). In addition to these effects, NPY has also been shown to promote remarkable effects on some aspects of tumor progression, including cell proliferation, matrix invasion, metastasis, and angiogenesis. All of these effects are mediated by specific NPY receptors (NPY-Rs), which belong to the rhodopsin-like G-protein-coupled receptor superfamily and currently encompass five cloned members in mammals, the Y1-, Y2-, Y4-, Y5- and y6-R. NPY-R activation, through the coupling to G_i, affects multiple response-specific second messenger cascades, such as the decrease of forskolin-stimulated cAMP accumulation in some different cell lines and tissues (Larhammar et al. 1992; Mullins et al. 2002; Pedrazzini et al. 2003), the inhibition of adenylyl cyclase (Michel et al. 1998), and the induction of Ca^{++} release from intracellular stores via phospholipase C (Gerald et al. 1995; Herzog et al. 1992). Moreover, NPY-R stimulation has been associated to the phosphorylation of the extracellularly regulated kinases (ERK/1/2) subgroup of the mitogen activated protein kinase (MAPK), dependently or not on protein kinase C (PKC) activity (Hansel et al. 2001; Mullins et al. 2002). The ERK/1/2 pathway, especially associated to the Y1-R and Y2-R (Nie, and Selbie 1998), seems to mediate the proliferative or antiproliferative effects of NPY in different cell systems, like primary cardiomyocytes (Pellieux et al. 2000), human erythroleukemia cells (Keffel et al. 1999), neuronal cells (Goldberg et al. 1998), human prostate cancer (PCa) cells (Ruscica et al. 2006). Evidence exists that the NPY system may play a role in the normal prostate as well as in the progression of human PCa, which represents one of the most common malignant diseases among men in the Western world (Abrahamsson 1999). In the human prostate, particularly in the smooth muscle layer, NPY is mainly localized in nerve fibers and neuroendocrine cells (Tainio 1995; Ventura et al. 2002). In addition, immunopositivity for NPY has been shown in 75% of the PCa specimens obtained from a series of patients (Mack et al. 1997), suggesting a possible participation of this factor in PCa growth and progression. Moreover, a recent study conducted on 243 human prostatectomy specimens has confirmed a role of NPY in the earliest stages of PCa, showing that aberrant expression of NPY is associated with biochemical recurrence after treatment for localized PCa (Rasiah et al. 2006).

2 NPY System in Prostate Cancer

With the aim of a better understanding of the role of NPY in PCa biology, our studies (Magni and Motta 2001; Ruscica et al. 2006), conducted on the three different human PCa cell lines LNCaP (androgen-dependent), DU145 and PC3 (androgen-independent), indicate that the Y1-R gene and protein are expressed in all these clones (Magni et al. 2001; Ruscica et al. 2006), while only PC3 cells express Y2-R. Y4-R expression has been found only in DU145 and LNCaP cells, while NPY and Y5-R are not expressed in any cell line (personal observation). Exposure of PCa cells to NPY reduced the proliferation of LNCaP and DU145 cells, whereas it stimulated that of PC3 cells. The maximally effective concentration of NPY (about 10^{-8} M) for each clone is in agreement with the expected range for this peptide (Reubi et al. 2001), as well as with the affinity of NPY to binding sites present on PC3 cell membranes (Kd: 30 nM). It should be noted that this Kd value is only slightly lower than that found for SK-N-MC cells (used in this study as a positive control) by Gordon (Gordon et al. 1990) and us (Magni et al. 2001). It has been reported that NPY can inhibit the growth of the Y1-expressing SK-N-MC cell line and that Y1-R has been identified in proliferating tissues such as primary and metastatic breast carcinoma (Reubi et al. 2001). NPY actions on PCa cells appear quite specific for cell proliferation, since this peptide has been reported to have no effect on the in vitro invasion and migration of PC3 (Nagakawa et al. 1998), LNCaP and DU145 cells (Nagakawa et al. 2001). The effect of NPY on cell growth in the human PCa cell lines appears mediated by Y1-R activation, since a pre-treatment with the selective Y1-R antagonist BIBP3226 abolishes this action (Ruscica et al. 2006).

The study of different transduction mechanisms associated with Y1-R activation and the consequent proliferative or antiproliferative effect led us to observe that in LNCaP cells no apparent involvement of ERK1/2 is present, similarly to other cell lines (Gallego et al. 1992; Pumiglia and Decker 1997). On the contrary, in both androgen-independent clones, DU145 and PC3, NPY treatment rapidly stimulates ERK1/2 phosphorylation, which quickly returns to the low basal levels in PC3 cells, but shows a persistent elevation (at least up to 6 h) in DU145 cells (Ruscica et al. 2006). These data agree with previous studies conducted in different cancers or immortalized cells, showing that a long-lasting stimulation of this pathway resulted in a reduced proliferation rate, as we observed in DU145 cells, whereas rapid and transient activation of MAPK leads to enhanced cell growth (as found in PC3 cells) (Alblas et al. 1998; Kim et al. 2003; Mabuchi et al. 2004). Furthermore, NPY-induced ERK1/2 phosphorylation appears mediated by an upstream activation of PKC only in PC3 cells, whereas in DU145 cells this process is clearly PKC-independent. It is interesting to note that a pretreatment with BIBP3226 abolishes Y1-R-mediated ERK1/2 activation. Moreover, NPY treatment was found to inhibit forskolin-induced cAMP accumulation only in PC3, but not in LNCaP and DU145 cells. Differently from neural cells expressing NPY-Rs, exposure to NPY did not affect $[Ca^{2+}]_i$ in any PCa cell line tested (Ruscica et al. 2006).

3 Concluding Remarks

In conclusion, the data reported in this paper indicate that the NPY system might have a pathophysiological significance in the context of PCa progression. According to our observations, NPY is not produced in vitro by the human PCa cell lines tested. NPY has, however, been found in the context of both human normal (Tainio 1995) and tumoral prostate (Mack et al. 1997; Rasiah et al. 2006). Therefore, in the prostate, NPY might be released both by neuroendocrine cells and nerve terminals to reach NPY-Rs localized nearby epithelial cells of normal or tumoral origin.

The evidence that NPY and its related membrane receptors are overexpressed in PCa, together with a large variety of human neoplasms (Reubi 1997), highlights the pathophysiological, diagnostic and therapeutical relevance of this neuroendocrine system, which needs to be further exploited in future studies. NPY receptors may represent a useful target for imaging and radiotherapy by radiolabeled peptide hormone analogos, similarly to what has been proposed for other neuroendocrine systems, like somatostatin and its analogs (de Herder et al. 2005).

Acknowledgments. This work was supported in part by grants from the University of Milan and the Ministero dell'Istruzione, Università e Ricerca (FIRB 2001, No. RBNE01JKLF_005).

References

Abrahamsson, P. A. (1999) Neuroendocrine cells in tumor growth of the prostate. Endocr. Relat. Cancer 6, 503–519.
Alblas, J., Slager-Davidov, R., Steenbergh, P. H., Sussenbach, J. S. and van der Burg, B. (1998) The role of MAP kinase in TPA-mediated cell cycle arrest of human breast cancer cells. Oncogene 16, 131–139.
de Herder, W. W., Kwekkeboom, D. J., Valkema, R., Feelders, R. A., van Aken, M. O., Lamberts, S. W., van der Lely, A. J. and Krenning, E. P. (2005) Neuroendocrine tumors and somatostatin: imaging techniques. J. Endocrinol. Invest. 28, 132–136.
Gallego, C., Gupta, S. K., Heasley, L. E., Qian, N. X. and Johnson, G. L. (1992) Mitogen-activated protein kinase activation resulting from selective oncogene expression in NIH 3T3 and rat 1a cells. Proc. Natl. Acad. Sci. U. S. A. 89, 7355–7359.
Gerald, C., Walker, M. W., Vaysse, P. J., He, C., Branchek, T. A. and Weinshank, R. L. (1995) Expression cloning and pharmacological characterization of a human hippocampal neuropeptide Y/peptide YY Y2 receptor subtype. J. Biol. Chem. 270, 26758–26761.
Goldberg, Y., Taimor, G., Piper, H. M. and Schluter, K. D. (1998) Intracellular signaling leads to the hypertrophic effect of neuropeptide Y. Am. J. Physiol. 275, C1207–1215.
Gordon, E. A., Kohout, T. A. and Fishman, P. H. (1990) Characterization of functional neuropeptide Y receptors in a human neuroblastoma cell line. J. Neurochem. 506–513.
Hansel, D. E., Eipper, B. A. and Ronnett, G. V. (2001) Neuropeptide Y functions as a neuroproliferative factor. Nature 410, 940–944.
Herzog, H., Hort, Y. J., Ball, H. J., Hayes, G., Shine, J. and Selbie, L. A. (1992) Cloned human neuropeptide Y receptor couples to two different second messenger systems. Proc. Natl. Acad. Sci. U. S. A. 89, 5794–5798.
Keffel, S., Schmidt, M., Bischoff, A. and Michel, M. C. (1999) Neuropeptide-Y stimulation of extracellular signal-regulated kinases in human erythroleukemia cells. J. Pharmacol. Exp. Ther. 291, 1172–1178.
Kim, D. S., Park, S. H., Kim, S. E., Kwon, S. B., Park, E. S., Youn, S. W. and Park, K. C. (2003) Lysophosphatidic acid inhibits melanocyte proliferation via cell cycle arrest. Arch. Pharm. Res. 26, 1055–1060.
Larhammar, D., Blomqvist, A. G., Yee, F., Jazin, E., Yoo, H. and Wahlestedt, C. (1992) Cloning and functional expression of a human neuropeptide Y/peptide YY receptor of the Y1 type. J. Biol. Chem. 267, 10935–10938.
Mabuchi, S., Ohmichi, M., Kimura, A., Ikebuchi, Y., Hisamoto, K., Arimoto-Ishida, E., Nishio, Y., Takahashi, K., Tasaka, K. and Murata, Y. (2004) Tamoxifen inhibits cell proliferation via mitogen-activated protein kinase cascades in human ovarian cancer cell lines in a manner not dependent on the expression of estrogen receptor or the sensitivity to cisplatin. Endocrinology 145, 1302–1313.

Mack, D., Hacker, G. W., Hauser-Kronberger, C., Frick, J. and Dietze, O. (1997) Vasoactive intestinal polypeptide (VIP) and neuropeptide tyrosine (NPY) in prostate carcinoma. E. J. Cancer 33, 317–318.

Magni, P. (2003) Hormonal control of the neuropeptide Y system. Curr. Protein Pept. Sci. 4, 45–57.

Magni, P. and Motta, M. (2001) Expression of neuropeptide Y receptors in human prostate cancer cells. Ann. Oncol. 12, S27–S29.

Michel, M. C., Beck-Sickinger, A., Cox, H., Doods, H. N., Herzog, H., Larhammar, D., Quirion, R., Schwartz, T. and Westfall, T. (1998) XVL International Union of Pharmacology recommendations for the nomenclature of neuropeptide Y, peptide YY and pancreatic polypeptide receptors. Pharmacol. Rev. 50, 143–150.

Mullins, D. E., Zhang, X. and Hawes, B. E. (2002) Activation of extracellular signal regulated protein kinase by neuropeptide Y and pancreatic polypeptide in CHO cells expressing the NPY Y(1), Y(2), Y(4) and Y(5) receptor subtypes. Regul. Pept. 105, 65–73.

Nagakawa, O., Ogasawara, M., Murata, J., Fuse, H. and Saiki, I. (2001) Effect of prostatic neuropeptides on migration of prostate cancer cell lines. Int. J. Urol. 8, 65–70.

Nagakawa, O., Ogasawara, M., Fujii, H., Murakami, K., Murata, J., fuse, H. and Saiki, I. (1998) Effect of prostatic neuropeptides on invasion and migration of PC-3 prostate cancer cells. Cancer Lett. 133, 27–33.

Nie, M. and Selbie, L. A. (1998) Neuropeptide Y Y1 and Y2 receptor-mediated stimulation of mitogen-activated protein kinase activity. Regul Pept 75-76, 207-213.

Pedrazzini, T., Pralong, F. and Grouzmann, E. (2003) Neuropeptide Y: the universal soldier. Cell. Mol. Life Sci. 60, 350–377.

Pellieux, C., Sauthier, T., Domenighetti, A., Marsh, D. J., Palmiter, R. D., Brunner, H. R. and Pedrazzini, T. (2000) Neuropeptide Y (NPY) potentiates phenylephrine-induced mitogen-activated protein kinase activation in primary cardiomyocytes via NPY Y5 receptors. Proc. Natl. Acad. Sci. U. S. A. 97, 1595–1600.

Pumiglia, K. M. and Decker, S. J. (1997) Cell cycle arrest mediated by the MEK/mitogen-activated protein kinase pathway. Proc. Natl. Acad. Sci. U. S. A. 94, 448–452.

Rasiah, K. K., Kench, J. G., Gardiner-Garden, M., Biankin, A. V., Golovsky, D., Brenner, P. C., Kooner, R., O'Neill G, F., Turner, J. J., Delprado, W., Lee, C. S., Brown, D. A., Breit, S. N., Grygiel, J. J., Horvath, L. G., Stricker, P. D., Sutherland, R. L. and Henshall, S. M. (2006) Aberrant neuropeptide Y and macrophage inhibitory cytokine-1 expression are early events in prostate cancer development and are associated with poor prognosis. Cancer Epidemiol. Biomark. Prev. 15, 711–716.

Reubi, J. C. (1997) Regulatory peptide receptors as molecular targets for cancer diagnosis and therapy. Q. J. Nucl. Med. 41, 63–70.

Reubi, J. C., Gugger, M., Waser, B. and Schaer, J. C. (2001) Y1-Mediated Effect of Neuropeptide Y in Cancer: Breast Carcinomas as Targets. Cancer Res. 61, 4636–4641.

Ruscica, M., Dozio, E., Boghossian, S., Bovo, G., Martos Riano, V., Motta, M. and Magni, P. (2006) Activation of the Y1 receptor by neuropeptide y regulates the growth of prostate cancer cells. Endocrinology 147, 1466–1473.

Tainio, H. (1995) Peptidergic innervation of the human prostate, seminal vesicle and vas deferens. Acta Histochem. 97, 113–119.

Ventura, S., Pennefather, J. N. and Mitchelson, F. (2002) Cholinergic innervation and function in the prostate gland. Pharmacol. Ther. 94, 93–112.

Chapter 8
Short Abstracts – Session II

Early Epigenetic and Genetic Events in Carcinogenesis

Thea D. Tlsty

Departments of Pathology, Surgery, Laboratory Medicine*, and UCSF Comprehensive Cancer Center, University of California at San Francisco, San Francisco, CA 94143-0511, USA.

One of the major problems with treating cancer is that it is detected too late. Early lesions and transitions often occur 10–15 years prior to our ability to detect the disease by palpation or imaging. By the time many cancers are detected they have long since acquired the ability to generate micrometastases or populations of tumor cells that can develop drug resistance. Our best approach to treating or preventing cancer is early detection and an understanding of the earliest steps in the transition of normal cells to malignant ones. However, there are considerable obstacles to studying the early steps of cancer formation in vivo. For one, the multitude of random genetic aberrations that are generated during the carcinogenic process mask the specific nonrandom and mechanistically related aberrations that would provide insights into how the lesion is initiated. Finding the critical alterations in a sea of changes is difficult. Furthermore, the pre-malignant lesions are small, sporadic and arise in a nonsynchronized fashion, making detection difficult and analysis almost impossible. Finally, at the present time, these cells have not been successfully cultured in vitro. Our ignorance about the earliest molecular changes prevents studies that would identify individuals at risk or identify targets for prevention strategies. An in vitro model that mimics the transitions that occur as normal epithelial cells become malignant would provide major advances in several aspects of dealing with the disease. Early detection of lesions could be used for risk assessment and diagnosis of the disease at early stages. Additionally, and perhaps, the most significant advance would be the insights a model system may provide to the initiation of lesions and the transition of these lesions to malignancy allowing the translation of that information into preventive agents. It is for these reasons that we began studying human epithelial cells in vitro, focusing on mammary epithelial cells in particular. Recently, studies of human epithelial cells and fibroblasts from healthy individuals have been providing novel insights into how early epigenetic and genetic events affect genomic integrity and fuel carcinogenesis. Key epigenetic changes, such as the hypermethylation of the p16 promoter sequence, create a previously unappreciated pre-clonal phase of tumorigenesis in which a subpopulation of epithelial cells is positioned for progression to malignancy (Nature 409:636, 2001). These key changes generate epigenetic and genetic mosaicism, precede the clonal outgrowth of pre-malignant lesions and occur frequently in healthy, disease-free individuals (Caner Cell 5:263, 2004). Prior work from our laboratory has identified biomarkers that may be useful for risk assessment as well as provide targets for the elimination of these cells. Understanding more about these early events should provide novel molecular candidates for prevention and therapy of cancer.

Romanov, S, Krystyna Kozakiewicz, Charles R. Holst, Martha R. Stampfer, Larisa M. Haupt, and Tlsty TD "Normal Human Mammary Epithelial Cells Spontaneously Emerge from Senescence and Acquire Genomic Instability" Nature, 2001; 409:633–637.

Crawford, Y. Gauthier, M., Joubel, A., Kozakiewicz, K., and Tlsty TD "Histologically Normal Human Mammary Epithelia with Silenced p16INK4a Over-express COX-2, Promoting a Premalignant Program" Cancer Cell 5 2004; pp. 263–273.

Tumor Promotion by Tumor-Associated Macrophages

Alberto Mantovani, MD

Istituto Clinico Humanitas, Rozzano and Università degli Studi di Milano.

Recent years have seen a renaissance of the inflammation-cancer connection stemming from different lines of work and leading to a generally accepted paradigm (1–4). An inflammatory component is present in the microenvironment of most neoplastic tissues, including those not causally related to an obvious inflammatory process. Cancer-associated inflammation includes: the infiltration of white blood cells, prominently phagocytic cells called macrophages (TAM) (5); the presence of polypeptide messengers of inflammation (cytokines such as tumor necrosis factor (TNF) or interleukin-1 (IL-1), chemokines such as CCL2); the occurrence of tissue remodeling and angiogenesis. Chemokines have emerged as a key component of the tumor microenvironment which shape leukocyte recruitment and function as well as tumor cell biology (6). Strong direct evidence suggests that cancer associated inflammation promotes tumor growth and progression. Therapeutic targeting of cancer promoting inflammatory reactions is in its infancy, and its development is crucially dependent on defining the underlying cellular and molecular mechanisms in relevant systems.

Balkwill F, Charles KA, Mantovani A. Smoldering and polarized inflammation in the initiation and promotion of malignant disease. Cancer Cell 2005; 7: 211–217.
Balkwill F, Mantovani A. Inflammation and cancer: back to Virchow? Lancet 2001; 357: 539–545.
Coussens LM, Werb Z. Inflammation and cancer. Nature 2002; 420: 860–867.
Mantovani A. Cancer: inflammation by remote control. Nature 2005; 435: 752–753.
Mantovani A, Sozzani S, Locati M, Allavena P, Sica A. Macrophage polarization: tumor-associated macrophages as a paradigm for polarized M2 mononuclear phagocytes. Trends Immunol 2002; 23: 549–555.
Mantovani A, ed. Chemokines in Neoplastic Progression. Sem Cancer Biol. Vol. 14. 2004.

Tumor Invasion in the Absence of Epithelial-Mesenchymal Transition: Podoplanin-mediated Remodeling Actin Cytoskeleton

Andreas Wicki*, Francois Lehembre*, Nikolaus Wick[§], Brigitte Hantusch[§], Dontscho Kerjaschki[§], and Gerhard Christofori*

*Institute of Biochemistry and Genetics, DKBW, Department of Clinical-Biological Sciences, Center of Biomedicine, University of Basel, Switzerland
[§]Clinical Institute for Pathology, Medical University of Vienna, Vienna

Tumor cell invasion into the surrounding tissue can exhibit a phenotype of either single cell or collective cell migration. Whereas single cell migration is mainly dependent on signaling pathways within migrating cells themselves and is usually accompanied by a loss of E-cadherin expression in many experimental cancer models, collective cell migration requires the maintenance of cell–cell adhesions and an organization of the tumor tissue. We have show that the expression of podoplanin, a small mucin-like protein, is upregulated in a number of human carcinomas. We have investigated podoplanin function in cultured human breast cancer cells, in a mouse model of pancreatic b cell carcinogenesis, and in human cancer biopsies. Here, we report a novel molecular pathway of tumor progression that does not involve the loss of E-cadherin function or epithelial-mesenchymal transition. Podoplanin prompts tumor cell spreading, migration and invasion in vitro and in vivo without dissolving E-cadherin-mediated junctional complexes. Furthermore, it induces actin cytoskeleton rearrangement and the formation of filopodia by modulating the activities of Rho-family GTPases, which ultimately leads to collective cell invasion, a phenotype often observed in human carcinomas. In conclusion, podoplanin mediates an alternative pathway of tumor cell invasion in the absence of epithelial-mesenchymal transition (EMT).

Section 3
Animal Models

Chapter 9
The *Arf* Tumor Suppressor in Acute Leukemias: Insights from Mouse Models of Bcr–Abl-Induced Acute Lymphoblastic Leukemia

Richard T. Williams[1] and Charles J. Sherr[2]

[1] Department of Oncology, St. Jude Children's Research Hospital, USA, richard.williams@stjude.org
[2] Department of Genetics and Tumor Cell Biology, St. Jude Children's Research Hospital, Howard Hughes Medical Institute, USA, sherr@stjude.org

Abstract. The prototypical Bcr–Abl chimeric oncoprotein is central to the pathogenesis of chronic myelogenous leukemias (CMLs) and a subset of acute lymphoblastic leukemias (Ph+ ALLs). The constitutive tyrosine kinase transforms either hematopoietic stem cells (in CML) or committed pre-B lymphoid progenitors (in Ph+ ALL) to generate these distinct diseases. The *INK4A/ARF* tumor suppressor locus is frequently deleted in both B- and T-lineage ALLs, including Ph+ ALL, whereas the locus remains intact in CML. In murine bone marrow transplant models and after transfer of syngeneic Bcr–Abl-transformed pre-B cells into immunocompetent recipient animals, *Arf* gene inactivation dramatically decreases the latency and enhances the aggressiveness of Bcr–Abl-induced lymphoblastic leukemia. Targeted inhibition of the Bcr–Abl kinase with imatinib provides highly effective therapy for CML, but Ph+ ALL patients do not experience durable remissions. Despite exquisite in vitro sensitivity of *Arf*-null, BCR–ABL+ pre-B cells to imatinib, these cells efficiently establish lethal leukemias when introduced into immunocompetent mice that receive continuous, maximal imatinib therapy. Bcr–Abl confers interleukin-7 (IL-7) independence to pre-B cells, but imatinib treatment restores the requirement for this cytokine. Hence, IL-7 can reduce the sensitivity of Bcr–Abl+ pre-B cells to imatinib. Selective inhibitors of both Bcr–Abl and the IL-7 transducing JAK kinases may therefore prove beneficial in treating Ph+ ALL.

1 Organization and Regulation of the *INK4A/ARF* Locus

The CDKN2A tumor suppressor locus (hereafter *INK4A/ARF*) encodes two distinct tumor suppressor genes (Lowe and Sherr 2003). The first, p16^{INK4A}, functions as an inhibitor of the cyclin D-dependent kinases, and restricts RB/E2F-dependent transcriptional programs required for entry into S phase. The second gene product, p14^{ARF} (p19^{Arf} in the mouse), binds and inactivates HDM2 (Mdm2 in the mouse), an E3 ubiquitin ligase that negatively regulates the p53 tumor suppressor. While ordinarily repressed during much of development and in most normal tissues, *Arf* is induced in response to abnormally elevated and sustained mitogenic signals that stem from oncogene activation. Upon *Arf* induction, a p53-dependent transcriptional response ensues that culminates in either cell cycle arrest or apoptosis depending on cellular context.

The *INK4A* and *ARF* genes are transcribed from distinct, alternate first exons but share common second and third exons that are translated in alternate reading frames (hence the designation ARF for the second gene characterized at this locus). Another member of the *INK4* subfamily of cyclin-dependent kinase inhibitors, *INK4B*, is located in close proximity to *INK4A–ARF*. Hence, this genomic region (*INK4B—ARF–INK4A*) encodes negative regulators of both RB and p53, allowing inactivation of both tumor suppressor pathways through a single deletional event. Deletion of the entire locus is frequently detected in a variety of cancers, including ALL.

2 BCR–ABL Induces Myeloid and Lymphoid Leukemias

The Philadelphia chromosome (Ph) (Nowell and Hungerford 1960) formed by the reciprocal translocation involving human chromosomes 9 and 22 (Rowley 1973) represents the founding genetic lesion in almost all cases of CML and in a subset of pediatric and adult ALL. The characteristic t(9;22)(q34;q21) translocation fuses the breakpoint cluster region (BCR) from chromosome 22 to a portion of the c-*Abl* proto-oncogene, producing alternative chimeric Bcr–Abl oncoproteins p185$^{BCR-ABL}$ and p210$^{BCR-ABL}$ (hereafter p185 and p210) that characterize CML and Ph+ ALLs, respectively (Chan et al. 1987; Clark et al. 1987; Groffen et al. 1984). Both oncoproteins exhibit constitutive tyrosine kinase activity essential for cellular transformation (Clark et al. 1987; Witte et al. 1980). While CML results from aberrant p210 expression in hematopoietic stem cells (HSCs) and multilineage and committed progenitors, p185 appears to be restricted to the lymphoid lineage in Ph+ ALL (Chan et al. 1987).

The targeted Abl tyrosine kinase inhibitor, imatinib, has proven remarkably successful in inducing durable remissions in most patients with chronic phase CML (Deininger et al. 2005). However, CML patients maintained on continuous drug therapy are not cured, and about 5% per year develop overt imatinib resistance, most frequently due to the evolution of kinase domain mutations in Bcr–Abl that interfere with imatinib binding (Shah et al. 2002). Second generation Abl kinase inhibitors, such as dasatinib and nilotinib, have now demonstrated pre-clinical and clinical activity against almost all previously identified mutant Bcr–Abl isoforms (Talpaz et al. 2006; Kantarjian et al. 2006). These inhibitors also target a broader spectrum of tyrosine kinases, which might independently contribute to their increased clinical effectiveness.

In Ph+ ALL patients, imatinib therapy often induces relatively brief remissions, and tragically these patients have poor prognoses despite conventional combination chemotherapy and myeloablative bone marrow transplantation. Like most other cases of pediatric ALLs (Pui et al. 2004), these Ph+ leukemias often have sustained deletions at chromosome 9p21 that encompasses the *INK4A/ARF* locus. In contrast, myeloblasts from CML patients in chronic phase (where imatinib is most effective), and in the clinically aggressive accelerated and blastic phases of the disease (where imatinib is less efficacious), display neither genomic loss at the *INK4A/ARF* locus nor *ARF* promoter methylation (our unpublished observations). We therefore wondered whether *INK4a/ARF* loss might contribute to imatinib resistance in Ph+ ALL.

3 Modeling Bcr–Abl-Induced Leukemias

Initial attempts to model BCR–ABL-induced leukemias focused on producing valid murine representations of human CML. Early approaches with either transgenic and retroviral bone marrow transduction/transplantation methodologies produced multiple hematopoietic neoplasms including CML, ALL, erythroid leukemias, T cell lymphoma and macrophage tumors (Wong and Witte 2001). However, through the optimization of the transplantation approach with 5-fluorouracil pretreatment of donor animals, tailored cytokine stimulation of cultured progenitors, and more efficient retroviral transduction of murine hematopoietic stem cells (HSCs) with BCR–ABL-expressing vectors, valid CML models emerged (Pear, et al. 1998; Zhang and Ren 1998). In contrast, Bcr–Abl transduction of a subset of the abundant lymphoid progenitors present in freshly isolated, nonconditioned whole bone marrow results in a highly penetrant pre-B cell lymphoblastic leukemia. The Src kinase family members Lyn, Hck, and Fgr are activated by Bcr-Abl, and engineered deletions of any two of them significantly impair pre-B cell leukemogenesis but do not impact on the efficiency of Bcr–Abl-induced myeloid disease (Hu et al. 2004).

4 *Arf* Function in Mouse Models of BCR–ABL-Induced Acute Lymphoblastic Leukemia

4.1 *BCR–ABL Activates the Arf Checkpoint*

Primary, murine pre-B cells derived from bone marrow progenitors can be readily established on a pre-existing stromal cell layer that secretes interleukin-7 (IL-7). Although these cells have a limited lifespan in culture and ultimately senesce, their *Arf*-null (or *p53*-null) counterparts, while still IL-7-dependent, can proliferate indefinitely (Randle et al. 2001). In mice, the *Ink4a* gene does not contribute to pre-B cell senescence. Abl oncoproteins confer IL-7-independence, but the oncogene-mediated activation of the *Arf-p53* checkpoint in pre-B cells triggers apoptosis and thereby limits their outgrowth. When *Arf* or *p53* are inactivated, the latter restraint is removed and transformed cells emerge.

When infected with retroviral vectors expressing Bcr–Abl isoforms together with green fluorescent protein (GFP), stromally supported cultures of infected bone marrow cells generated from *Arf* $^{+/+}$ (Wild-type, WT) and *Arf* $^{-/-}$ (*Arf*-null) mice each yield homogenous pre-B cell populations (>95% GFP+) coexpressing either the p185 or p210 isoforms. However, upon transfer to stroma-free liquid culture conditions with no supporting cytokines, Bcr–Abl-expressing WT cells rapidly undergo apoptosis, whereas *Arf*-null p185+ or *Arf*-null p210+ cells proliferate exponentially in an IL-7-independent manner (Williams et al. 2006). Pre-B cells derived from *Arf* $^{+/-}$ (heterozygous) animals display an intermediate phenotype, and their "adaptation" to growth in liquid culture correlates with inactivation of the remaining WT *Arf* allele, behavior characteristic of a classic "two-hit" tumor suppressor gene. Thus, expression of Bcr–Abl in the absence of *Arf* allays the oncogene-mediated activation of the p53 transcriptional program and facilitates the robust proliferation of pre-B cells without any requirement for exogenous IL-7.

4.2 *Arf Inactivation in a Mouse Bone Marrow Transplantation Model of Ph+ ALL*

Utilizing a conventional bone marrow transplantation (BMT) approach to model BCR–ABL-induced ALL (Hu et al. 2004), unconditioned donor bone marrow, transduced with retroviruses coexpressing Bcr–Abl and GFP, was transplanted without intervening in vitro culture into lethally irradiated recipient mice. Mice that received transduced *Arf* $^{+/+}$ (WT) donor cells all succumbed to a lethal lymphoblastic leukemia with a median survival of 5–6 weeks. Analysis of recipients of transduced WT donor cells confirmed Bcr–Abl-dependent expression of p19^{Arf} in circulating leukocytes at three weeks post-transplantation. Robust Bcr–Abl-dependent p19^{Arf} and p53 induction was also detected in isolated bone marrow and splenic B-lineage cells recovered from clinically well mice as early as two weeks after transplantation of transduced WT donor cells. In contrast, all recipients of transduced *Arf* $^{-/-}$ bone marrow cells developed a more aggressive leukemia and survived less than three weeks post-transplant (Williams et al. 2006). Immature pre-B leukemic cells (with a B220+ CD24+, BP-1 +, sIgM – immunophenotype) infiltrated the spleen and bone marrow of diseased mice, but greater organ infiltration and dramatically elevated numbers of circulating GFP+ lymphoblasts were observed in recipients of transduced *Arf* $^{-/-}$ donor cells. These experiments emphasize the effectiveness of the *Arf*-dependent checkpoint in vivo to restrain (at least temporarily) the rapid evolution of a lethal lymphoid leukemia. Bcr–Abl-transduced p53$^{-/-}$ donor cells initiate a very aggressive lymphoid leukemia indistinguishable from that generated by Bcr–Abl transduction of *Arf* $^{-/-}$ donor cells (our unpublished observations).

Together, these results suggest that, in this experimental system, much of *Arf*'s potent tumor suppressive function is p53-dependent.

4.3 Arf Inactivation Contributes to Aggressive Disease in Immunocompetent Mice

A methodology that complements the conventional BMT approach entails generation of primary Bcr–Abl-expressing murine pre-B cells of pre-determined genotypes (as described in Sect. 4.1) and directly measuring their leukomogenic potential upon transfer to syngeneic C57BL/6 recipient mice. Initial experiments employed intra-peritoneal (IP) injection of pre-B cells into nonirradiated, immunocompetent recipient animals. IP injection of 2×10^6 $Arf^{-/-}$ p210+ cells rapidly and universally induced an aggressive lympholeukemia, whereas similar numbers of $Arf^{+/+}$ p210+ cells rarely produced detectable disease. Heterozygous $Arf^{+/-}$ p210+ cells produced a partially penetrant lympholeukemia, consistent with their relative proliferative advantage in liquid culture.

A more refined approach entails intravenous (IV) injection of serial log dilutions of p210+ or p185+ donor cells into recipient mice. While 2×10^5 $Arf^{+/+}$ p210+ cells failed to establish leukemias in recipient animals and 2×10^6 $Arf^{+/+}$ p185+ cells produced disease with a 4–5 week latency, as few as 200 $Arf^{-/-}$ p210+ cells or 20 $Arf^{-/-}$ p185+ cells (the lowest numbers yet evaluated) were capable of producing an aggressive leukemia within four weeks of injection (our unpublished data). These experiments argue that short-term *Arf*-null, p185+ and *Arf*-null p210+ pre-B cell cultures are highly enriched in leukemia-initiating cell activity, and that there may be little requirement (or opportunity) for in vivo selection of other collaborating genetic events. Furthermore, as five logs fewer *Arf*-null p185+ cells than p185+ WT cells were required to initiate leukemia in recipient animals, these experiments underscore the exceptional potency of *Arf* as a tumor suppressor in vivo.

Additional benefits arise from the use and further refinement of this syngeneic pre-B cell transfer model. First, the recipient mice require no pre-conditioning or hematopoietic reconstitution, retain their normal immune function and resist opportunistic infections. Second, initiating pre-B cells and their leukemic progeny recovered from moribund mice can be further cultured ex vivo, genetically and biologically analyzed and compared, and can be serially transplanted into secondary host animals. Finally, large cohorts of recipient mice can be inoculated with fixed numbers of genetically programmed cells and then subjected to therapeutic trials.

4.4 Imatinib Resistance in Murine BCR–ABL+, Arf-Null ALL

In vitro, *Arf*-null p185+ and *Arf*-null p210+ pre-B cells are exquisitely sensitive to growth inhibition by the tyrosine kinase inhibitor imatinib (IC$_{50}$ ~ 100 nM). Like their *p53*-null counterparts, they undergo cell cycle arrest at low concentrations of drug (<1 µM) but rapidly undergo apoptosis in response to sustained exposure to low micromolar concentrations. Dose-dependent inhibition of tyrosine phosphorylation of the p185 and p210 kinases themselves and of bona fide downstream substrates (e.g., Stat5, CrkL) correlates with growth inhibition and occurs independently of *Arf* status (Williams et al. 2006).

However, *Arf*-null p210+ pre-B cells introduced IV into recipient animals produced fatal leukemias despite continuous treatment of the mice with high dose oral imatinib [100 mg kg^{-1} twice daily, capable of inducing durable remissions in murine CML models (Wolff and Ilaria 2001) and comparable to maximal human therapy]. Imatinib treatment extended median survival of host animals by ~7 days, and reduced spleen weights and circulating lymphoblast counts, confirming a measurable biological response. Importantly, leukemic cells recovered from moribund, imatininb-treated mice retained their original in vitro imatinib sensitivity, arguing that drug resistance was not due to amplification of, or mutations in, the Bcr–Abl kinase, but rather to cell extrinsic host factors that protected the leukemic cells from drug-induced cytostasis.

While cytokine-independence is a useful surrogate, if not requirement, for full transformation by oncogenes, it does not preclude cytokine-responsiveness of transformed cells per se. Addition of IL-7 to *Arf*-null Bcr–Abl+ pre-B cells partially counteracted the growth-inhibitory effects of imatinib, even at therapeutically desirable and -achievable low micromolar drug concentrations, and correlated with preservation of cyclin D2 expression and continued cell proliferation. Janus (JAK) kinases are critical elements in cytokine common gamma chain (γ_c)-dependent signaling, including the response to IL-7. As would be expected, targeting JAK kinases with a small molecule inhibitor resensitized leukemic cells to imatinib even in the presence of saturating levels of IL-7 (Williams et al. 2006). Whereas the nontumor cell-autonomous basis of imatinib resistance of Bcr–Abl+, *Arf*-null pre-B cells in our ALL model could potentially reflect any number of host-dependent protective mechanisms, the latter experiments provide proof-of-principle that imatinib sensitivity can be influenced by cytokines.

5 Conclusions

In primary murine pre-B cells, the Bcr–Abl oncogene confers IL-7-independence, efficiently induces *Arf,* and enforces a robust *Arf*- and *p53*-dependent apoptotic response that curtails their further expansion. In contrast, immortal *Arf*-null pre-B cells transformed by p185 or p210 are refractory to these cell death-promoting signals. Thus, the combination of Bcr–Abl expression and *Arf* inactivation provide complementary qualities of factor-independence, resistance to oncogene-induced apoptosis, and cellular immortality. With a conventional transplantation approach, we demonstrated potent in vivo *Arf*-dependent tumor suppression; *Arf* inactivation significantly decreased tumor latency and increased the severity of the lymphoblastic leukemia. In a syngeneic pre-B cell transfer model, Bcr–Abl-dependent, cell-autonomous *Arf* tumor suppression provides highly robust protection against leukemia development. Together, these studies raise the intriguing possibility that for vigorous lymphoid leukemogenesis, there may be little requirement for additional genetic abnormalities other than Bcr–Abl expression and *Arf* inactivation.

Despite exquisite imatinib sensitivity in vitro, highly leukemogenic *Arf*-null p210+ pre-B cells display cell-extrinsic imatinib resistance in vivo. Imatinib-treated animals succumbed to disease just days after vehicle-treated control animals died, and while their leukemic burdens in spleen and the peripheral circulation were reduced by therapy, bone marrow infiltration was equally significant. This is consistent with the idea that host-derived factors (e.g., cytokines like IL-7 produced in a normal host's bone marrow microenvironment) can counter growth inhibition by imatinib in vivo. *Arf* status does not appear to directly affect imatinib-induced inhibition of the Bcr–Abl signaling pathway. However, we reason that at drug concentrations that inhibit the Bcr–Abl kinase, leukemic cells would be resensitized to cytokines, and their survival might therefore be rescued by interleukins (or analogous factors) available within the bone marrow microenvironment. Restoration of imatinib responsiveness with JAK kinase inhibition in vitro paves the way for a rigorous assessment of cytokine-dependent imatinib resistance in vivo.

The highly enriched leukemia-initiating activity within the *Arf*-null, Bcr–Abl+ pre-B cell cultures will facilitate testing of the next generation Bcr–Abl kinase inhibitors, provides a platform for biological characterization, genomic analysis and genetic manipulation of leukemia-initiating cells, and validates the use of this versatile model system for the development of novel therapeutic strategies to control this refractory leukemia.

Acknowledgments. We thank the members of the Sherr and Roussel laboratory for helpful suggestions, criticisms and assistance with various aspects of this project. This work was supported in part by National Institutes of Health Training Grant T32-CA70089 (R.T.W.) and the American Lebanese Syrian Associated Charities of St. Jude Children's Research Hospital. C.J.S. is an Investigator of the Howard Hughes Medical Institute.

References

Chan, L.C., Karhi, K.K., Rayter, S.I., Heisterkamp, N., Eridani, S., Powles, R., Lawler, S.D., Groffen, J., Faulkes, J.G., Greaves, M.F., and Wiedemann, L.M. (1987) A novel *abl* protein expressed in Philadelphia chromosome positive acute lymphoblastic leukemia. Nature 325, 635–637.

Clark, S.S., McLaughlin, J., Crist, W.M., Champlin, R., and Witte, O.N. (1987) Unique forms of the *abl* tyrosine kinase distinguish Ph[1]-positive CML from Ph[1]-positive ALL. Science 235, 85–88.

Deininger, M., Buchdunger, E., and Druker, B.J. (2005) The development of imatinib as a therapeutic agent for chronic myelogenous leukemia. Blood 105, 2640–2653.

Groffen, J., Heisterkamp, N., de Klein, A., Bartram, C.R., and Grosveld, G. (1984) Philadelphia chromosomal breakpoints are clustered within a limited region, *bcr*, on chromosome 22. Cell 36, 93–99.

Hu, Y., Liu, Y., Pelletier, S., Buchdunger, E., Warmuth, M., Fabbro, D., Hallek, M., Van Etten, R.A., and Li, S. (2004) Requirement of Src kinases Lyn, Hck, and Fgr for BCR–ABL1-induced B-lymphoblastic leukemia but not chronic myeloid leukemia. Nat. Genet. 36, 453–461.

Kantarjian, H., Giles, F., Wunderle, L., Bhalla, K., O'Brien, S., Wassmann, B., Tanaka, C., Manley, P., Rae, P., Mietlowski, W., Bochinski, K., Hochhaus, A., Griffin, J.D., Hoelzer, D., Albitar, M., Dugan, M., Cortes, J., Alland, L., and Ottmann, O.G. (2006) Nilotinib in imatinib-resistant CML and Philadelphia chromosome-positive ALL. New Engl. J. Med. 354, 2542–2551.

Lowe, S.W. and Sherr, C.J. (2003) Tumor suppression by *Ink4a–Arf*: progress and puzzles. Curr. Opin. Genet. Dev. 13, 77–83.

Nowell, P. and Hungerford, D. (1960) Chromosomes of normal and leukemic human leukocytes. J. Natl. Cancer Inst. 25, 85.

Pear, W.S., Miller, J.P., Xu, L., Pui, J.C., Soffer, B., Quackenbush, R.C., Pendergast, A.M., Bronson, R., Aster, J.C., Scott, M.L., and Baltimore, D. (1998) Efficient and rapid induction of a chronic myelogenous leukemia-like myeloproliferative disease in mice receiving P210 bcr/abl-transduced bone marrow. Blood 92, 3780–3792.

Pui, C.-H., Relling, M.V., and Downing, J.R. (2004) Acute lymphocytic leukemia. New Engl. J. Med. 350, 1535-1548.

Randle, D.H., Zindy, F., Sherr, C.J., and Roussel, M.F. (2001) Differential effects of $p19^{Arf}$ and $p16^{Ink4a}$ loss on senescence of murine bone marrow-derived pre-B cells and macrophages. Proc. Natl. Acad. Sci. USA 98, 9654–9659.

Rowley, J.D. (1973) A new consistent chromosomal abnormality in chronic myelogenous leukemia. Nature 243, 290–293.

Shah, N.P., Nicoll, J.M., Nagar, B., Gorre, M.E., Paquette, R.L., Kuriyan, J., and Sawyers, C.L. (2002) Multiple BCR–ABL kinase domain mutations confer polyclonal resistance to the tyrosine kinase inhibitor imatinib (STI571) in chronic phase and blast crisis chronic myeloid leukemia. Cancer Cell 2, 99–102.

Talpaz, M., Shah, N.P., Kantarjian, H., Donato, N., Nicoll, J., Paquette, R., Cortes, J., O'Brien, S., Nicaise, C., Bleickardt, E., Blackwood-Chirchir, M.A., Iyer, V., Chen, T.T., Huang, F., Decillis, A.P., and Sawyers, C.L. (2006) Dasatinib in imatinib-resistant Philadelphia chromosome-positive leukemias. New Engl. J. Med. 354, 2531–2541.

Williams, R.T., Roussel, M.F., and Sherr, C.J. (2006) *Arf* gene loss enhances oncogenicity and limits imatinib response in mouse models of Bcr–Abl-induced acute lymphoblastic leukemia. Proc. Natl. Acad. Sci. USA 103, 6688–6693.

Witte, O.N., Dasgupta, A., and Baltimore, D. (1980) Abelson murine leukemia virus protein is phosphorylated *in vitro* to form phosphotyrosine. Nature 283, 826–831.

Wolff, N.C. and Ilaria, R.L. (2001) Establishment of a murine model for therapy-related chronic myelogenous leukemia using the tyrosine kinase inhibitor STI571. Blood 98, 2808–2816.

Wong, S. and Witte, O.N. (2001) Modeling Philadelphia chromosome positive leukemias. Oncogene 20, 5644–5659.

Zhang, X. and Ren, R. (1998) Bcr–Abl efficiently induces a myeloproliferative disease and production of excess interleukin-3 and granulocyte-macrophage colony-stimulating factor in mice: a novel model for chronic myelogenous leukemia. Blood 92, 3829–3840.

Chapter 10
Short Abstracts – Session III

Mouse Models to Understand the Role of Telomeres and Telomerase in Cancer and Aging

Maria A. Blasco

Telomeres and Telomerase Group, Molecular Oncology Program, Spanish National Cancer Centre (CNIO), Madrid, Spain

Telomeres protect the chromosome ends from unscheduled DNA repair and degradation. Telomeres are heterochromatic domains composed of repetitive DNA (TTAGGG repeats) bound to an array of specialized proteins. The length of telomere repeats and the integrity of telomere-binding proteins are both important for telomere protection. In addition, we have recently shown that telomere length is regulated by a number of epigenetic modifications, thus pointing to a higher-order control of telomere function. A key process in organ homeostasis is the mobilization of stem cells out of their niches. Defects in organ homeostasis are present both in cancer and in aging-related diseases. Here we will discuss that telomere length and the catalytic component of telomerase, Tert, are critical determinants for the mobilization of epidermal stem cells. On one hand, we will show that telomere shortening in the absence of telomerase negatively impacts on the mobilization of epidermal stem cells. On the other hand, Tert over-expression in the absence of changes in telomere length significantly increases the mobilization of epidermal stem cells, thus providing a mechanism by which Tert may promote tumorigenesis independently of telomere length. Finally, we will describe the generation and characterization of mice with constitutive expression of the telomere-binding protein TRF2 in the skin. TRF2 mice show a remarkable phenotype in the skin consisting of hyper-pigmentation, hair loss, dry skin, as well as increased skin tumors, all of which are reminiscent of the skin abnormalities characteristic of Xeroderma pigmentosum (XP) syndrome. We propose that the XP-like skin phenotypes described here for TRF2 mice are the result of a combination of defective DNA repair together with short telomeres, thus pinpointing to the roles of TRF2 in the context of the organism. In addition, this new mouse model demonstrates the impact of altered TRF2 expression both on cancer and aging.

MicroRNA Expression Signature Predicts Lung Cancer Diagnosis and Prognosis

Carlo M. Croce

The John Wolfe Professor for Cancer Research, The Ohio State University, Comprehensive Cancer Center, Columbus, OH 43210, USA

MicroRNA (miRNA) expression profiles for lung cancers were examined to investigate miRNA's involvement in lung carcinogenesis. miRNA microarray analysis identified statistical unique profiles, which could discriminate lung cancers from no-cancerous lung tissues as well as molecular signatures that differ in tumor histology. miRNA expression profiles correlated with survival of lung adenocarcinomas, including those classified as disease stage I. High hsa-mir-155 and low hsa-let-7a-2 expression correlated with poor survival by univariate analysis as well as multivariate analysis for hsa-mir-155. The miRNA expression signature on outcome was confirmed by real-time RT-PCR analysis of precursor miRNAs and crossvalidated with an independent set of adenocarcinomas. These results indicate that miRNA expression profiles are diagnostic and prognostic markers of lung cancer.

Role of *c-fos* in Skin Differentiation and Tumor Formation

Juan Guinea Viniegra

Institute of Molecular Pathology, Erwin F. Wagner lab, Wien Austria
Juan Guinea Viniegra, Harald Scheuch, Rainer Zenz and Erwin F. Wagner
Institute of Molecular Pathology, Vienna

The proto-oncogene *c-fos* is a major nuclear target for signal transduction pathways involved in the regulation of cell growth, differentiation, and transformation. To investigate the function of *c-fos* in skin development and skin tumor formation, we achieved specific conditional deletion of *c-fos* in the epidermis. Mice lacking *c-fos* in the keratinocytes (*c-fos*) show normal skin and no obvious phenotype. In vitro treatment of c-fos f/f or *c-fos* keratinocytes with Ca^{2+} or with the promoting agent, TPA, induced premature differentiation of the keratinocytes in the absence of *c-fos*. A similar phenotype was observed in newborn and adult mice treated topically with TPA. The observed premature keratinocyte differentiation in the *c-fos* mice is due to an increased Notch1 activation, which induces an increase in p21 and Caspase 3 protein expression. In the context of oncogenic signals driven by H-RasV12, *c-fos* keratinocytes overexpressing Ras show again premature differentiation. In the absence of *c-fos*, tumor-prone K5-SOS transgenic mice show strikingly reduced papilloma formation. Further analysis showed that the tumors consist of highly differentiated cells, while no differences in apoptosis or proliferation were observed. Interestingly, conditional and inducible deletion of *c-fos* in the tumor prone K5-SOS transgenic mice, led to a shutdown in tumor growth. These tumors again exhibited a highly differentiated state. In summary, premature activation of Notch1 in *c-fos* deficient keratinocytes leads to increased differentiation resulting in reduced tumor formation.

Oncogene-Induced DNA Damage Response and Tumor Suppression in eu-MYC Transgenic Mice: Critical Role of Tip60

ChiaraGorrini*, Massimo Squatrito*, Samantha Bennett, Domenico Sardella, John Lough1, and Bruno Amati

European Institute of Oncology, Milan, Italy and 1Medical College of Wisconsin, Milwaukee, WI 53226, USA.
*These authors have contributed equally to this work.

Oncogene activation is thought to induce a DNA-damage response (DDR) that activates p53 and suppresses tumor progression. The acetyl-transferase Tip60 is a cofactor for Myc and p53 but also has a direct function in DDR. We addressed the role of Tip60, in Myc-induced DDR and lymphomagenesis by crossing Eu-myc transgenic mice with heterozygous Tip60 knockout animals. The B-cells of young Eu-myc mice, but not of nontransgenic littermates, showed a sustained DDR as judged by phosphorylation of ATM, H2AX, Chk1 and p53 (Ser 15). Tip60 heterozygosity severely impaired this DDR and caused a two-fold acceleration in lymphomagenesis, without significantly affecting pre-tumoral B-cell proliferation or apoptosis. Lymphomas arising in Eu-myc p53$^{+/-}$ mice always lose the remaining p53 allele (LOH), while mutations that impinge on p53 activity prevent LOH. The selective pressure to loose p53 was retained in Eu-myc p53$^{+/-}$ Tip60$^{+/-}$ mice, then Tip60 haplo-insufficiency does not bypass p53 function. Thus, at least in Eu-myc mice, the DDR is redundant (most likely with the ARF pathway) for p53 activation and suppresses tumorigenesis through additional routes. In summary, oncogenic activation of Myc in B-cells causes a sustained DDR in vivo and a mutation hampering this DDR – here Tip60 heterozygosity – accelerates tumor progression.

Section 4
Transcription and Epigenetics

Chapter 11
Short Abstracts – Session IV

How the Epigenome Changes in Cancer

Peter A. Jones

USC/Norris Comprehensive Cancer Center, 1441 Eastlake Avenue, Los Angeles, CA 90089, USA

It is becoming increasingly clear that epigenetic silencing of tumor suppressor genes plays a causative role in human carcinogenesis. Epigenetic silencing involves changes in DNA cytosine methylation as well as chromatin structural changes which act to reinforce gene inactivation and thus the progression of tumors. Of these processes, DNA methylation has been the most studied and it has been convincingly shown that de novo methylation of CpG islands in the promoter regions of genes is linked to their heritable silencing. It has also become clear that histone modifications including the deacetylation of histones and the application of specific histone marks are associated with this process as is the binding of methylated DNA binding proteins such as MeCP2. Chromatin remodeling is becoming more frequently linked to such inactivation as well. We have recently showed that several CpG island promoters are free of nucleosomes in actively expressing cells and that nucleosome occupancy is associated with the silencing discussed above. Reversing chromatin structural changes including DNA methylation, histone modification and nucleosome occupancy is an important therapeutic target for epigenetic therapy. Since epigenetic changes are potentially reversible, the role of epigenetic therapy in cancer treatment is likely to become more relevant over the next few years.

Role of Ubiquitin-Like Proteins in the Control of Gene Expression

Ronald T. Hay

Centre for Interdisciplinary Research, School of Life Sciences, University of Dundee, Scotland, UK

The activity and subnuclear localization of many nuclear proteins and transcription factors are modulated by attachment of the small ubiquitin-like modifier SUMO. SUMO-1 and its relatives SUMO-2 and SUMO-3 are covalently linked to target proteins by a specific conjugation pathway involving SAE1/SAE2 (E1), Ubc9 (E2) and E3 ligases. SUMO modification is highly dynamic and the extent of SUMO modification is a balance between conjugation and deconjugation. A family of SUMO specific proteases is responsible not only for deconjuagtion, but also for processing the SUMO precursor. We have isolated and characterized a series of these proteases from human cells. Structural analysis of the SUMO protease SENP1 has revealed the mechanism of substrate recognition and proteolytic cleavage. In vivo SUMO conjugation has diverse roles, but modification of transcription factors often results in transcriptional repression. SUMO specific proteases can derepress the transcription factor indicating that a balance between SUMO conjugation and deconjugation is an important determinant of gene expression. During the physiological response of cells to heat shock this balance is shifted and SUMO modification of many substrates is increased. Quantitative proteomics has allowed us to identify hundreds of SUMO substrates and track changes in the extent of SUMO modification during these responses.

Regulation of Genomic Repression and its Link to Cancer

Ramin Shiekhattar

Associate Professor, Molecular Genetics Program, The Wistar Institute Philadelphia, USA

RNA interference is implemented through the action of a multiprotein complex termed RNA-induced silencing complex (RISC) programmed through microRNA or siRNA. We describe the biogenesis of microRNA mediated by the Microprocessor complex. We will also present evidence supporting a role for the RISC as a pre-assembled stable multiprotein complex composed of Dicer, the double-stranded RNA binding protein TRBP, and Ago2. We demonstrate that this complex could cleave a target RNA using a precursor microRNA (pre-miRNA) hairpin as the source of siRNA. Moreover, the Dicer-containing RISC could distinguish the guide strand of the siRNA from the passenger strand and specifically incorporates only the guide strand into an active RISC. Importantly, while ATP hydrolysis is not required for miRNA processing, assembly of an active RISC, or target RNA cleavage, ATP stimulates target RNA cleavage yielding enhanced RISC activity. These results define the composition of the RISC, and demonstrate that miRNA processing by Dicer and Ago2-mediated target cleavage are coupled.

The IKK Complex: Providing a Link between Inflammation and Cancer

Michael Karin, Ph.D.

Laboratory of Gene Regulation and Signal Transduction, Department of Pharmacology, University of California, San Diego, School of Medicine, La Jolla, CA 92093-0723, USA

A link between inflammation and cancer has been suspected for over two millennia, but its molecular nature remained ill defined. It has also been observed that certain bacterial (for instance *Helicobacter pylori*) and viral (for instance HBV and HCV) pathogens are major risk factors for certain types of cancer, most notably gastric and liver cancers. In trying to understand molecular mechanisms that link chronic infections and inflammation to cancer, we have postulated that transcription factor NF-κB may be at the center of this nexus, as NF-κB is activated in response to infection and inflammation and in turn upregulates expression of antiapoptotic and growth promoting genes. As there are several NF-κB transcription factors, we decided to inactivate the critical catalytic subunit of the IκB kinase (IKK) complex, IKKβ, as a way to inhibit activation of most NF-κB forms. We used conditional gene targeting to inactivate IKKβ in either cells that give rise to the malignant component of the tumor or in myeloid cells that contribute to the inflammatory infiltrate present in most tumors. Using a mouse model of colitis-associated cancer (CAC), we found that although deletion of the IKKβ subunit of the IKK complex in intestinal epithelial cells does not decrease inflammation, it does lead to a dramatic decrease in tumor incidence without affecting tumor size. This effect was linked to increased epithelial apoptosis during tumor promotion. A more modest reduction in tumor number but a considerable decrease in tumor size is caused by deletion of IKKβ in myeloid cells. This deletion diminishes expression of pro-inflammatory cytokines, COX-2 and MMP-9, without affecting apoptosis. These results show that the IKK/NF-κB pathway is not only involved in suppression of apoptosis in advanced cancers but that its specific inactivation in two different cell types, one of which plays a bystander role, can attenuate formation of inflammation-associated tumors. Thus, IKKβ may provide a mechanistic link between inflammation and cancer. Using a different model, based on transplantation of a syngeneic colon carcinoma cell line (CTC26) into immunocompetent mice, we have found that the IKK/NF-κB pathway is involved in inflammation-induced tumor progression and metastatic growth. In this case, inhibition of NF-κB activation in the cancer cell converted inflammation-induced tumor growth to inflammation-induced tumor regression. Importantly, we found that inflammation promoted tumor growth through the induction of TNF-α, which activated NF-κB in the cancer cell. In addition to inhibition of inflammation-induced proliferation, blocking NF-κB in the cancer cell greatly increased sensitivity to TRAIL-induced apoptosis. While inflammation is a major factor that contributes to the development and progression of CAC and other inflammation-linked cancers and is estimated to be involved in up to 20% of all human cancers, we asked whether inflammation driven by NF-κB has an important role in other forms of cancer where chronic inflammation or infection do not precede tumor development. To that end, we used a model of chemically-induced hepatocellular carcinoma (HCC) based on exposure of mice to a complete and potent carcinogen – diethyl nitrosamine (DEN). Heretofore, DEN administration, although resulting in pronounced cytotoxicity, was not found to trigger an inflammation response. Surprisingly, mice lacking IKKβ only in hepatocytes (IkkβΔhep mice) exhibited a marked increase in hepatocarcinogenesis after DEN administration. This increase correlated with enhanced reactive oxygen species (ROS) production, increased JNK activation and elevated hepatocyte death, giving rise to augmented compensatory proliferation of surviving hepatocytes. Brief oral administration of an anti-oxidant around the time of DEN exposure blocked prolonged JNK activation and compensatory proliferation and prevented excessive DEN-induced carcinogenesis in IkkβΔhep mice. A similar decrease in compensatory proliferation and hepatocarcinogenesis was observed in response to a knockout of the Jnk1 locus. Decreased hepatocarcinogenesis was also found in mice lacking IKKβ in both hepatocytes and hematopoietic-derived Kupffer cells. These mice exhibited reduced hepatocyte regeneration and diminished induction of hepatocyte growth factors, which were unaltered in IkkβΔhep mice. IKKβ, therefore, orchestrates inflammatory crosstalk between hepatocytes and hematopoietic-derived cells that promotes chemical hepatocarcinogenesis. Most likely, this inflammatory response is triggered by proteins released by hepatocytes undergoing necrosis in response to DEN administration and cased activation of nearby Kupffer cells, which in turn secrete growth factors that stimulate the proliferation of surviving hepatocytes including those that acquired DEN-induced oncogenic mutation. This inflammatory response to cellular necrosis may stimulate the growth and progression of many solid cancers.

References

Karin, M., Cao, Y., Greten, F. R. & Li, Z. W. NF-κB in cancer: from innocent bystander to major culprit. Nat Rev Cancer 2, 301–310 (2002).

Greten, F. R. et al. IKKβ links inflammation and tumorigenesis in a mouse model of colitis-associated cancer. Cell 118, 285–296 (2004).

Luo, J. L., Maeda, S., Hsu, L. C., Yagita, H. & Karin, M. Inhibition of NF-kappaB in cancer cells converts inflammation-induced tumor growth mediated by TNFalpha to TRAIL-mediated tumor regression. Cancer Cell 6, 297–305 (2004).

Maeda, S., Kamata, H., Luo, J. L., Leffert, H. & Karin, M. IKKβ couples hepatocyte death to cytokine-driven compensatory proliferation that promotes chemical hepatocarcinogenesis. Cell 121, 977–990 (2005).

Greten, F. R. & Karin, M. NF-κB: Linking Inflammation and Immunity to Cancer Development and Progression. Nat Rev Immunol (2005).

Section 5
High-Throughput Approaches and Imaging

Chapter 12
Identification and Validation of the Anaplastic Large Cell Lymphoma Signature

Roberto Piva,* Elisa Pellegrino, and Giorgio Inghirami

Department of Pathology and Center for Experimental Research and Medical Studies (CeRMS), University of Torino, Italy,
*roberto.piva@unito.it

Abstract. Anaplastic large cell lymphomas (ALCL) represent a subset of lymphomas in which the anaplastic lymphoma kinase (*ALK*) gene is fused to several partners, most frequently to the *NPM* gene. We have previously demonstrated that the constitutive expression and phosphorylation of ALK chimeric proteins is sufficient for cellular transformation, and its activity is strictly required for the survival of ALCL cells. To unravel signaling pathways required for NPM–ALK-mediated transformation and tumor maintenance, we analyzed the transcriptomes of ALK positive ALCL cell lines through experimentally controlled approaches in which ALK signaling was abrogated by an inducible ALKshRNA or by ALK inhibitors. Transcripts derived from the gene expression profiling analyses uncovered a reproducible signature, which includes a novel group of ALK-regulated genes. A functional RNAi screening identified new ALK transcriptional targets instrumental to cell transformation and/or to sustain the growth and survival of ALK positive ALCL cells. Thus, we prove that an experimentally controlled and functionally validated gene expression profiling analysis represents a powerful tool to identify novel pathogenetic networks and to validate biologically suitable target genes for therapeutic interventions.

While this paper was in press, a more detailed version of the study herein described was published by the same authors.
Piva, R., Pellegrino, E., Mattioli, M., Agnelli, L., Lombardi, L., Boccalatte, F., Costa, G., Ruggeri, B.A., Cheng, M., Chiarle, R., Palestro, G., Neri, A., and Inghirami, G "Functional Validation of the Anaplastic Lymphoma Kinase Signature Identifies CEBPB and Bcl2A1 as Critical Target Genes" J Clin Invest 116:3171-3182, 2006.

1 Introduction

The development of an oncogenic state is a complex process occurring through a range of defects in cell-signaling pathways, which allow cancer cells to alter their normal programs of proliferation, transcription, growth, migration, differentiation, and death. The ability to define different neoplastic subsets, recurrence of disease, and response to specific therapies based on the deregulation of specific cell signaling pathways has been demonstrated in multiple studies by the advent of microarray-based gene expression signatures. These efforts have been mainly concentrated on unbiased screening of cancer transcriptomes and have yielded, so far, important insights (Ebert and Golub 2004; Staudt and Dave 2005). However, these approaches have to face the enormous genetic heterogeneity within tumor samples. Therefore, these expression signatures might result in relatively "not reproducible" profiles, once applied on a wider scale for diagnostic purposes. Furthermore, unbiased approaches are bound to identify the end-points of complex cancer-causing alterations, in which the pathogenetic events are not distinguishable from transcriptional conditions consequential to proliferation, stage of differentiation, etc. These problems can be circumvented by exploiting highly controlled experimental approaches, relying on the modulation of oncogene(s) expression and by monitoring the effects by microarray. In this regard, various studies have demonstrated the potential for using gene expression profiles for the analysis of oncogenic pathways (Shaffer et al., 2006). However, other caveats affecting experimental gene expression profile studies should be considered. For example, gain-of-function experiments suffer from the fact that the cellular context for oncogene expression is often inappropriate (i.e., HeLa cells or mouse embryonal fibroblast cannot be representative of the oncogenic signaling in hematological malignancies). On the other hand, loss-of-function experiments (whether by dominant-negative constructs, RNA interference, or pharmacologic treatment) might suffer from potential lack of specificity (Jackson et al., 2003). Another significant problem of microarray experiments is linked to their capability of generating long lists of genes, without providing precise clues on the causal events responsible for a given phenotype. For these reasons, a careful experimental design is critical and it is important that samples for comparison are as closely matched as possible. These issues can be successfully reduced by inducible systems and well-designed time course experiments. Once a set of genes has been identified as candidates for explaining a tumoral phenotype, their function has to be modulated artificially to determine their degree of involvement in a given process. The combined use of gene expression profiling and RNA interference technology has been demonstrated a valuable strategy to characterize the biological functions of several genes and validate potential drug targets (Smith et al., 2006). Importantly, RNAi has been recently developed as a means to probe gene function on a whole-genome scale in mammalian cells (Kolfschoten et al., 2005; Westbrook et al., 2005).

In this study we identified and validated new gene targets involved in lymphomagenesis and in tumor maintenance utilizing as a model anaplastic large cell lymphoma (ALCL) cells.

Anaplastic large cell lymphomas (ALCL) represent a subset of non-Hodgkin lymphomas (NHLs) expressing CD30 and characterized by specific chromosome translocations in which the anaplastic lymphoma kinase (*ALK*) gene is fused to several partners, most frequently to the *NPM* gene (Kutok and Aster 2002; Pulford et al., 2004). The *ALK* gene encodes a tyrosine kinase receptor whose physiologic expression in mammalians is largely limited in specific regions of the central and the peripheral nervous system. ALK fusion proteins maintain the catalytic domain of ALK receptor at their C-terminus, whereas the N-terminal region of each fusion protein contains a dimerization domain. As a consequence of dimerization, ALK chimeras undergo autophosphorylation and become constitutively active. We have previously demonstrated that the constitutive expression and phosphorylation of NPM–ALK is sufficient for cellular transformation in vitro and for the development of lymphoid neoplasia in transgenic mice (Chiarle et al., 2003). However, the precise mechanisms of NPM–ALK transformation and the requirements of ALK expression for tumor growth and survival are still unclear. Activated ALK chimeras bind multiple adaptor proteins capable of firing different pathways regulating cell proliferation, survival and cell transformation (Pulford et al. 2004). Using cell-lineage specific conditional knockout models, we have demonstrated that the genetic ablation of STAT3 in ALK positive cells leads to cell death, and prevents the generation of B cell neoplasms (Chiarle et al., 2005). Previously, PLC-γ and AKT have been shown to play an essential role in ALK-mediated transformation in vitro. It has been postulated that the activation of Ras/ERK and PLC-γ pathways contribute to enhancement of cell growth and that Stat3 and PI3K/AKT pathways play a major role in inhibiting apoptosis. Nevertheless, it is conceivable that ALK-mediated transformation requires a more complex scenario.

To clarify the precise mechanisms and signaling pathways required for ALK-mediated transformation and tumor growth, we analyzed the transcriptomes of ALK positive ALCL cell lines using tightly controlled experimental conditions, in which the ALK signaling was abrogated by inducible ALK–shRNA or by potent ALK inhibitors. These approaches provided a "bona-fide" ALK oncogenic signature by filtering out the majority of off-target effects triggered by each experimental system. Positive hits were functionally validated by RNAi to ascertain genes which could provide growth and survival benefits to tumor cells, both in cell culture and in animal models.

2 Generation of Inducible ALK knockdown Cell Lines

To identify candidate genes involved in the pathogenesis of ALCL, we expressed a doxycycline-inducible ALK–shRNA in multiple ALK positive cell lines. Transduction of controls and ALK–shRNA sequences was accomplished using lentiviral vectors. Doxycycline treatment determined a progressive down-modulation of NPM–ALK mRNA and protein expression (Figure 1A).

Conditional NPM–ALK knockdown was coupled to a G1 cell cycle arrest and was followed by a striking increase of apoptotic cells. To prove whether this system could abrogate NPM–ALK driven growth in vivo, we studied the growth patterns of xenograft tumors in immunocompromised mice. ALCL cells, injected subcutaneously, formed visible growing tumors after 2–3 weeks

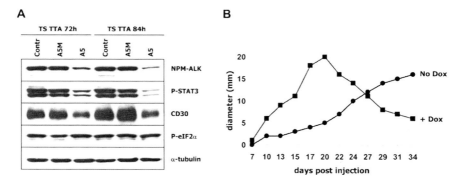

FIGURE 1. (**A**) Inducible NPM–ALK knockdown leads to down-modulation of ALK and of known downstream targets in TS–TTA–A5 cells. Protein expressions were assayed by Western blotting with the specified antibodies. (**B**) NPM–ALK silencing induces tumor regression in vivo. Representative tumor growth curves for TS–TTA–A5 cells injected subcutaneously into immunocompromised mice in an untreated mouse (*circles*) or a mouse treated for 14 days with Dox (*squares*) starting at day 17 mouse.

latency. In mice treated with doxycycline, tumor growth rapidly diminished and, after 10 days, most tumors were reabsorbed (Figure 1B). These findings confirm the strict requirement of ALK signaling for the sustained tumor growth and survival of ALK positive ALCLs (Piva et al. 2006). Moreover, inducible ALK–shRNA cells represent an ideal tool to untangle the *ALK* gene expression signature.

3 Identification of ALK Expression Signature in ALCL Cells

To identify reproducible signatures in multiple ALCL cell lines, we compared the gene expression profile of two ALCL cell lines, TS and Su-DHL1, prior to and after doxycycline-mediated ALK knockdown. Transcripts, whose expression is known to be regulated by NPM–ALK, such as CD30 (TNFRSF8), JUNB, MYC, and Survivin (BIRC5) were modulated as expected. On the contrary, the expression of classic interferon target genes (e.g., OAS1) was not significantly affected by RNAi (Figure 2).

To further validate the specificity of the NPM–ALK signature, we analyzed the expression profile upon induction of a mutated ALK–shRNA. The resulting NPM–ALK gene expression signature revealed a total of 345 and 149 probe-sets differentially expressed upon induction of ALK–shRNA in TS and Su-DHL1 cells, respectively. Comparison of both signatures showed

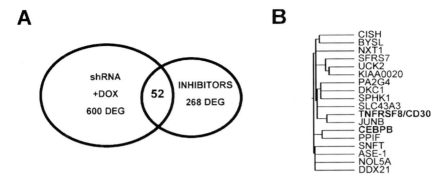

FIGURE 2. Generation of a restricted signature for ALK positive ALCL cells. (A) Overlapping transcripts concordantly modulated across shRNA- and ALK inhibitor-treated TS-TTA-A5 cells. (B) Genes strongly correlated to TNRSF8 expression. DEG: Differentially Expressed Genes.

that 69% of transcripts (103 genes) of Su-DHL1 were common to those of TS cells, with 72 increased and 31 decreased. Thus, we identified a consistent NPM–ALK signature common to ALCL cell lines.

To further validate the gene expression profile (GEP) signature obtained after RNAi and to exclude possible bias due to potential off-targets aberrantly modulated by ALK–A5 shRNA, we took advantage of potent cell permeable pyrrolocarbazole-derived ALK inhibitors (Wan et al. 2006). The supervised analysis of two ALK inhibitors-treated versus untreated and inactive compound-treated cells revealed a total of 268 differentially expressed transcripts. To identify a stronger and more reliable NPM–ALK signature, we then used two distinct approaches.

Overlap. We selected the overlapping set of genes identified by both ALK silencing and kinase inhibition gene expression analyses. A total of 52 common transcripts with 22 up-regulated and 30 down-regulated were identified. To provide evidence for the reliability of this combined NPM–ALK signature, we arbitrarily selected six gene targets (ICOS, RGS16, CCL20, DKC1, GNL3, BCL2A1) and verified that their transcripts were modulated accordingly to ALK activity.

Correlation. Since the mRNA expression of ALK was not changed by ALK inhibitors, we selected as a reference the tightly regulated NPM–ALK-induced gene *TNFRSF8* (CD30). The list of genes obtained from ALK inhibitor was then hierarchically clustered for correlation to TNRSF8 expression. The TNFRSF8 branch included many transcripts also annotated with the "overlap" approach such as JUNB, DKC1, and others, but also novel genes, including the transcription factor CCAAT/enhancer binding protein β (C/EBPβ) (Ramji and Foka 2002).

4 Functional Validation of ALK Signature

We reasoned that if any one of these genes acts as an oncogene and/or plays a relevant role in the maintenance of the ALCL neoplastic phenotype, its loss could be sufficient to affect cell survival, proliferation, or morphology, despite a sustained ALK activation. ALK transcriptional targets were validated through a small-scale genetic screening using lentiviral shRNA sequences. This screening allowed us to identify genes that promote the growth and survival of tumor cells, both in cell culture and in animal models. We provide an example of such validation by showing the pathogenetic role of C/EBPβ in ALK-driven transformation and survival.

To validate the microarray data, we carried out protein expression studies on ALCL cells and verified that the expression of C/EBPβ was strongly repressed following NPM–ALK inactivation in both TS and Su-DHL1 cells. To confirm these findings, we carried out Western blot analysis on a broad panel of T and B lymphoma/leukemia cell lines and found that C/EBPβ protein was specifically expressed in ALK positive ALCL cell lines. These data were confirmed in a panel of primary human lymphomas in which strong C/EBPβ expression was detected preferentially in ALK positive ALCL.

To assess the functional properties of C/EBPβ expression in ALCL cells, we selectively silenced its expression by lentiviral-mediated RNA interference. We found that C/EBPβ knockdown affected tumor cell viability in ALK positive cells, without any significant impairment of survival in the ALK negative cell lines. We subsequently investigated whether C/EBPβ was necessary for NPM–ALK-mediated cellular transformation of mouse embryonal fibroblasts (MEF). In vivo, $clebpb^{-/-}$ NPM–ALK MEFs generated significantly smaller tumors compared to wild-type MEFs when injected into athymic Nu/Nu mice, underlining the importance of C/EBPβ in sustaining the transforming properties of NPM–ALK. Taken together, our analyses demonstrate that C/EBPβ expression is restricted to ALK positive ALCLs and it is transcriptionally regulated by NPM–ALK activity. More importantly, C/EBPβ plays a crucial role in supporting both transformation and survival of NPM–ALK positive cells (Figure 3).

5 Conclusions

We have used gene expression profiling in association with a functional validation approach to identify the molecular mechanisms leading to NPM–ALK transformation and to discover feasible targets for specific ALCL therapies. To achieve this objective, we have adopted a loss-of-function approach with inducible lentiviral-mediated RNAi and potent ALK inhibitors in multiple human lymphoid cell lines transformed by NPM–ALK. This method produced a reliable and restricted expression signature of the oncogenic ALK in the appropriate cellular context, and it cross-validated both

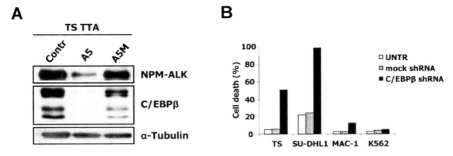

FIGURE 3. (A) C/EBPβ protein expression is strongly repressed following NPM–ALK silencing. TS–TTA cells were transduced with A5 or A5M ALK-shRNA and treated with DOX. Different bands represent multiple C/EBPβ isoforms. (B) C/EBPβ knockdown compromises viability of ALK positive ALCL cells. ALK positive ALCL (TS and Su-DHL1) and ALK negative (MAC-1 and K562) cell lines were transduced with C/EBPβ- or mock- shRNA constructs. Apoptotic cells were determined at day 7.

ALK inhibitors and shRNAs as viable therapeutic approaches for the treatment of ALK positive tumors. Using a battery of lentiviral shRNA constructs, we assessed the pathogenetic role of ALK regulated genes and their individual contribution in the generation and/or maintenance of ALK neoplastic phenotype. As a proof of principle, we showed that C/EBPβ expression is required to sustain the survival and growth of ALK positive ALCL cells.

In conclusion, this study demonstrates that by combining two of the most powerful new technologies in cancer research, RNAi and microarray analysis, it is possible to dissect the regulatory networks downstream of any oncogene within an appropriate cellular context. Furthermore, a systematic functional screening allows definition of mechanisms and identification of oncogenic pathways that can be therapeutically targeted in cancers.

Acknowledgments. This project was supported by NIH R01-CA64033, Sixth Research Framework Program of the European Union, Project RIGHT (LSHB-CT-2004-005276), Ministero dell'Università e Ricerca Scientifica (MIUR), Regione Piemonte, Compagnia di San Paolo, Torino (Progetto Oncologia), and Associazione Italiana per la Ricerca sul Cancro (AIRC). RP is supported by the MIUR program "Incentivazione alla mobilità di studiosi residenti all'estero".

References

Chiarle, R., Gong, J.Z., Guasparri, I., Pesci, A., Cai, J., Liu, J., Simmons, W.J., Dhall, G., Howes, J., Piva, R. and Inghirami, G. (2003) NPM–ALK transgenic mice spontaneously develop T-cell lymphomas and plasma cell tumors. *Blood*, 101, 1919–1927.

Chiarle, R., Simmons, W.J., Cai, H., Dhall, G., Zamo, A., Raz, R., Karras, J.G., Levy, D.E. and Inghirami, G. (2005) Stat3 is required for ALK-mediated lymphomagenesis and provides a possible therapeutic target. *Nat Med*, 11, 623–629.

Ebert, B.L. and Golub, T.R. (2004) Genomic approaches to hematologic malignancies. *Blood*, 104, 923–932.

Jackson, A.L., Bartz, S.R., Schelter, J., Kobayashi, S.V., Burchard, J., Mao, M., Li, B., Cavet, G. and Linsley, P.S. (2003) Expression profiling reveals off-target gene regulation by RNAi. *Nat Biotechnol*, 21, 635–637.

Kolfschoten, I.G., van Leeuwen, B., Berns, K., Mullenders, J., Beijersbergen, R.L., Bernards, R., Voorhoeve, P.M. and Agami, R. (2005) A Genetic Screen Identifies PITX1 as a Suppressor of RAS Activity and Tumorigenicity. *Cell*, 121, 849–858.

Kutok, J.L. and Aster, J.C. (2002) Molecular biology of anaplastic lymphoma kinase-positive anaplastic large-cell lymphoma. *J Clin Oncol*, 20, 3691–3702.

Piva, R., Chiarle, R., Manazza, A.D., Taulli, R., Simmons, W., Ambrogio, C., D'Escamard, V., Pellegrino, E., Ponzetto, C., Palestro, G. and Inghirami, G. (2006) Ablation of oncogenic ALK is a viable therapeutic approach for anaplastic large-cell lymphomas. *Blood*, 107, 689–697.

Pulford, K., Morris, S.W. and Turturro, F. (2004) Anaplastic lymphoma kinase proteins in growth control and cancer. *J Cell Physiol*, 199, 330–358.

Ramji, D.P. and Foka, P. (2002) CCAAT/enhancer-binding proteins: structure, function and regulation. *Biochem J*, 365, 561–575.

Shaffer, A.L., Wright, G., Yang, L., Powell, J., Ngo, V., Lamy, L., Lam, L.T., Davis, R.E. and Staudt, L.M. (2006) A library of gene expression signatures to illuminate normal and pathological lymphoid biology. *Immunol Rev*, 210, 67–85.

Smith, R., Owen, L.A., Trem, D.J., Wong, J.S., Whangbo, J.S., Golub, T.R. and Lessnick, S.L. (2006) Expression profiling of EWS/FLI identifies NKX2.2 as a critical target gene in Ewing's sarcoma. *Cancer Cell*, 9, 405–416.

Staudt, L.M. and Dave, S. (2005) The Biology of Human Lymphoid Malignancies Revealed by Gene Expression Profiling. Advances in Immunology. Academic Press, New York, pp. 163–208.

Wan, W., Albom, M.S., Lu, L., Quail, M.R., Becknell, N.C., Weinberg, L.R., Reddy, D.R., Holskin, B.P., Angeles, T.S., Underiner, T.L., Meyer, S.L., Hudkins, R.L., Dorsey, B.D., Ator, M.A., Ruggeri, B.A. and Cheng, M. (2006) Anaplastic lymphoma kinase activity is essential for the proliferation and survival of anaplastic large-cell lymphoma cells. *Blood*, 107, 1617–1623.

Westbrook, T.F., Martin, E.S., Schlabach, M.R., Leng, Y., Liang, A.C., Feng, B., Zhao, J.J., Roberts, T.M., Mandel, G., Hannon, G.J., Depinho, R.A., Chin, L. and Elledge, S.J. (2005) A Genetic Screen for Candidate Tumor Suppressors Identifies REST. *Cell*, 121, 837–848.

Chapter 13
Cell-Cycle Inhibitor Profiling by High-Content Analysis

Fabio Gasparri[1], Antonella Ciavolella[2] and Arturo Galvani[3]

[1]*Department of Biology, Nerviano Medical Sciences, Milano, Italy, fabio.gasparri@nervianoms.com*
[2]*Department of Biology, Nerviano Medical Sciences, Milano, Italy, antonella.ciavolella@nervianoms.com*
[3]*Department of Biology, Nerviano Medical Sciences, Milano, Italy, arturo.galvani@nervianoms.com*

Abstract. The discovery of agents which disrupt cancer cell division by specifically targeting key components of the cell-cycle machinery represents a major focus of recent drug discovery efforts in Oncology. The drug discovery process can be greatly enhanced by multiparametric cellular analysis which can assist in confirmation, often in a few multiplexed assays, of the mechanism of action (MOA) of compounds identified through biochemical screening or similar in vitro methods. High-Content Analysis (HCA) is a technique based on automated microscopy which enables multiparametric analysis of fluorescent indicators to define cellular responses to compound treatment. Several distinct fluorescence channels can be acquired and analyzed within a single measure in the same cell population. Here we present a multiparametric HCA approach to characterize potential cell-cycle inhibitors in osteosarcoma U-2 OS adherent cell cultures. This approach allows monitoring of compound-induced cell-cycle perturbations by analyzing specific cellular markers such as nuclear morphology, DNA content or histone H3 phosphorylation. Moreover, the induction of DNA damage response or apoptosis can also be readily evaluated. By considering the profile of the investigated cellular markers at different compound concentrations, a fingerprint defines the cellular and molecular phenotype associated with each compound.

High-content analysis (HCA) is a novel analytical approach which integrates fluorescence-based assays and image analysis algorithms for automated analysis of subcellular events and provides multidimensional single-cell phenotypic information (Giuliano et al. 1997; Giuliano et al. 2003). HCA allows quantitation at the single-cell level of fluorescence signal intensity associated with specific cellular markers evidenced by fluorescent dyes [e.g., DNA staining by 2-diamidino-5-phenylindole (DAPI)], immuno-fluorescent staining, or use of fluorescent proteins (Lang et al. 2006). Among the singular advantages offered by HCA with respect to other cell-based techniques, such as flow cytometry, is the possibility to determine the subcellular localization of fluorescent signals. This allows, for example, measurement of biological phenomena involving protein re-localization which are not necessarily accompanied by significant variation of overall expression levels (total fluorescence), such as nuclear translocation of transcription factors (Ding et al. 1998).

In addition to this, HCA can provide morphometric information regarding cell size, shape, neurite formation (Arden et al. 2002) or even the re-location of single live cells in the field during time frames (cell motility) (Richards et al. 2004). In particular, cellular processes inducing major changes in the nucleus/chromatin shape (mitosis, cytotoxicity, or apoptosis) can be followed by HCA. DNA staining with fluorescent dyes allows monitoring of specific nuclear morphological features (condensation, fragmentation, poly-nuclei formation) associated to these phenomena within a simple and low-cost measurement. Such an approach is noteworthy for the possibility to screen chemical libraries for discovery of cell-cycle inhibitors or apoptosis inducers through a simple analysis of the nuclear morphology (Smellie et al. 2006).

The acquisition and processing of distinct fluorescence channels by HCA within a single or few measurements of the same cell population enables multiplex analysis. Previous reports have described, for example, multiparameter apoptosis assays based on simultaneous analysis of caspase-3 activity, mitochondrial mass-potential and nuclear fragmentation/condensation (Lövborg et al. 2004). Another multiparameter application has been described by Minguez and co-workers, who analyzed four cellular markers (DNA content, α-tubulin, phospho-histone H3 and phospho-RSK90) in the same sample population (Minguez et al. 2002). Very recently, we presented a multiparameter HCA approach for cell-cycle analysis through simultaneous acquisition of four independent cell-cycle markers: DNA content, BrdU incorporation, cyclin B1 expression and histone H3 phosphorylation (Gasparri et al. 2006).

The application of HCA to large numbers of samples is referred to as "high-content screening" (HCS) (Giuliano et al., 1997). This medium-high throughput approach allows acquisition of detailed biological information on the activity of compound libraries in a cellular context, representing a fundamental advantage in the early evaluation of potential drugs. The relative slowness of the majority of HCS plate readers with respect to other instrumentations (e.g., ELISA readers), mainly due to autofocusing and long exposure times, limits the relative throughput and the applications of HCS to secondary cellular assays and compound mechanism of action (MOA) investigation. However, some of the more recent technical innovations introduced by manufacturers (such as faster autofocus or confocal-like optics), hint towards primary screening capabilities and to the possibility of conducting assays at subcellular resolution (Clemons 2004; Lang et al. 2006; Ramm 2005).

HCS has been already used in medium-high throughput assay formats, for example, to screen chemical libraries using kinetic and endpoint phenotypic assays in neural precursor cultures (Richards et al. 2006). Recently, Barabasz and co-workers characterized Aurora kinase inhibitors by measuring the levels of the mitotic marker phospho-histone H3 (Ser10) in treated cell cultures (Barabasz et al. 2006). Moreover, five potent centrosome-duplication inhibitors were identified screening a chemical library of about 16,000 compounds by an image-based centrosome-duplication assay based on centrosome count and size analysis (Perlman et al. 2005).

The majority of the HCS applications reported so far have employed one or two cellular markers to evaluate the biological activity of compounds in a specific pathway or cellular process of interest. The analysis of a series of markers representing distinct processes would allow generation of biological information on overall activity of compounds in a given cellular model: this is the concept of "high-content profiling" (HCP). Whereas classical cell-based screening can be considered a "direct approach" to drug discovery (only compounds active in a certain assay are selected), HCP corresponds to a "reverse approach": compounds previously selected on the basis of their generic cytotoxicity or cell-growth inhibition capability, for example, are tested in a set of cellular assays to identify their mechanism of action. Thus the main goal of HCP is the identification of the primary cellular target or the primary cellular mechanism with which the compound exerts its effect. Since a compound's activity depends on the dose and the time of treatment, its HCP profile should also be a function of both: dose–response titration and kinetic analysis allow investigation of potency and selectivity towards different targets, giving a "fingerprint" of the cellular and molecular phenotype associated with the drug.

One of the most exhaustive reports describing an MOA profiling of pharmacologically active compounds by image-based cellular assays was that of Perlman et al. (2004). These authors demonstrated the possibility of characterizing the cellular activity of a test set of 100 pharmacologically relevant drugs on different cellular markers. The set of markers analyzed was arbitrarily chosen and included morphology reporters, proliferation/cell cycle-related parameters, signal transduction markers and stress response markers. The authors treated a human cancer cell line (HeLa) with increasing doses of compounds for 20 h and acquired the fluorescence intensity distributions associated with each marker by automated microscopy. A dose-dependent phenotype profile was obtained for each compound through a statistical analysis [titration-invariant similarity score (TISS)] based on the maximum distance between the cumulative fluorescence distributions (Kolmogorov–Smirnov statistics) of treated cells at each compound dose and the cumulative distributions of untreated control cells. Following this strategy, the authors successfully performed hierarchical cluster analysis of compound profiles, observing clustering of molecules with known, similar MOAs, but of distinct chemical classes (such as nocodazole and podophyllotoxin, for example). Moreover, these authors were able to assess the cellular activity of poorly characterized compounds by profile similarity with known drugs. Among the advantages of this approach there is the possibility to compare distributions of arbitrary shape, the robustness towards variation in dynamic range and noise level and, most importantly, the insensitivity to the relationship between the fluorescence staining intensity and the antigen density. Nevertheless, this approach requires quite complex data analysis and it is intrinsically dependent on the quality of the untreated control populations. Moreover, the choice of the cellular markers to be analyzed, although made with the objective of investigating an overall representation of cellular processes, depends on the availability of high quality antibodies for the immunofluorescent detection of a given antigen, on the expression of that antigen in a certain cell line, and eventually can influence the outcome of the hierarchical clustering. This is one of the major problems in approaching a high-content compound profiling campaign: the list of cellular markers must be decided *a priori*, representing a bias in the evaluation of the activity of unknown compounds.

One of the leading technology platforms in the HCS field is the ArrayScan™ system (Cellomics Inc, Pittsburgh, USA), a fully automated inverted microscope which focuses samples and scans fields by means of a motorized stage. During acquisition, the fluorescence signals associated with up to six channels are simultaneously detected and quantified in distinct subcellular compartments (e.g., nucleus, cytoplasm, or membrane) to provide both morphometric and biochemical information related to each single cell. Quantitative fluorescence data can be managed by statistical analysis to provide segregated or aggregated information, for example, cell population analysis (Gasparri et al. 2004).

Here we describe the use of HCA to profile the mechanism of action of well characterized cell-cycle inhibitors with the ArrayScan™ system. The effects of 20 reference compounds with distinct MOAs on different cellular markers were analyzed in the osteosarcoma U-2 OS cell line. Compounds selected ranged from microtubule inhibitors (colchicine, nocodazole, paclitaxel, and vincristine), to inhibitors of the synthesis of DNA (gemcitabine and thymidine) or protein (anisomycin, puromycin, cyclohexymide). A full list of the compounds employed is reported in Table 1. The compound set that we assembled was chosen to cover common cellular mechanisms and/or targets relevant for cancer therapy (Gibbs 2000).

Cells were seeded at a density of 1,000/well in poly-Lys-coated 384-well plates (Matrix Technologies, Hudson, NH, USA) and treated with increasing compound concentrations (ranging between 128 pM and 10 µM; dilution factor 5) for 1 or 24 h. Fifty micromolar BrdU was added 15–20 min before fixation to all samples treated for 24 h. Seven cellular markers were chosen and quantified in cells by fluorescence staining to evaluate specific perturbations of the cell-cycle, as well as the induction of DNA damage or cell death (apoptosis or cytotoxicity):

– *BrdU incorporation*. Active S-phase marker.
– Cyclin B1 expression. G2/M-phase marker.
– Phospho-histone H3 (Ser 28). Mitotic marker.

TABLE 1. Reference standard compounds used in this study.

	Compound	Mechanism of action	Source (Cat.#)
1	Colchicine	Microtubules inhibitors	Biomol (T1180001)
2	Nocodazole		Sigma-Aldrich (M1404)
3	Paclitaxel		Sigma-Aldrich (T7402)
4	Vincristine		Biomol (T117-0005)
5	Gemcitabine	DNA synthesis inhibitors	Eli-Lilly
6	Thymidine		Sigma-Aldrich (T1895)
7	Anisomycin	Protein synthesis inhibitors	Sigma-Aldrich (A9789)
8	Cyclohexymide		Biomol (GR310-0100)
9	Puromycin		Biomol (GR312-0050)
10	MG-132	Proteasome inhibitors	Calbiochem (474790)
11	Bortezomib		Millennium
12	Camptothecin	DNA topoisomerase inhibitors	Sigma-Aldrich (C9911)
13	Etoposide		Sigma-Aldrich (E1383)
14	SN-38		Synthesized in house
15	Brostallicin	Minor groove DNA binder	Synthesized in house
16	Wortmannin	PI3K inhibitor	Sigma-Aldrich (W1628)
17	Rapamycin	mTOR inhibitor	Cell Signaling (9904)
18	Aurora inhibitor (compound 18, Fancelli et al. 2005)	Aurora inhibitor	Synthesized in house
19	CDKs inhibitor (compound 13, Pevarello et al. 2005)	CDKs inhibitor	Synthesized in house
20	Staurosporine	Kinase inhibitor	Sigma-Aldrich (S5921)

- Phospho-S6 ribosomal protein (Ser 235/236). Marker of mTOR activity
- Active caspase-3. Apoptosis marker.
- Nuclear area increase (DAPI staining). Nuclear fragmentation – S/G2 phase block – polynuclei formation.
- Nuclear area decrease (DAPI staining). Mitotic block – apoptotic nuclear condensation.

DAPI was added to each sample to allow single-cell recognition and nuclear morphology analysis. The cell number (average number of cells per field, expressed as % of remaining cells with respect to controls) was obtained by counting nuclei and was added to the profile to assess cytotoxicity and cell-cycle inhibition capability of compounds. We previously reported that the analysis of DNA content together with BrdU incorporation, cyclin B1 expression and histone H3 phosphorylation allows accurate estimation of the percentage of cells in the different cell-cycle phases and to follow cell-cycle perturbations due to compound treatment (Gasparri et al. 2006).

Cyclin B1 is a well-known G2/M marker: it starts to appear in late S-phase and accumulates in the cytoplasm during G2-phase (Sherr 1996). Histone H3 plays a key role in mitotic chromosome condensation: it is specifically phosphorylated during mitosis at Ser10 and Ser28 residues by the Aurora-B kinase (Hendzel i, et al. 1997). BrdU pulse labeling for short periods (15—20 min) labels S-phase cells actively engaged in DNA synthesis (Trent et al. 1986). We determined BrdU incorporation in cell cultures by a nondestructive approach described elsewhere based on BrdU immunostaining after nuclease treatment (Gonchoroff et al. 1986; Gasparri et al. 2006).

Phospho-S6 (Ser 235/236) is a marker of the mTOR/Akt pathway activation, thus it was included to monitor the inhibition of this important signaling pathway at early time points (1 h of treatment). Phosphorylation of the ribosomal protein S6 by p70S6K is implicated in protein synthesis regulation and eventually in coupling cell growth to cell-cycle progression upon growth factor simulation (Fingar et al. 2002).

Apoptosis is a highly regulated process of programmed cell death induced by internal factors (DNA damage, activation of genetic programs, etc.) or external factors (ionizing radiation, cytotoxic dugs, cytokines, etc.). The execution phase of apoptosis, mediated by activated execution caspases, such as caspase-3 (Cohen 1997), is characterized by marked changes in cell morphology that include cytoskeletal modifications, contraction and cellular membrane blebbing, that eventually result in the formation of apoptotic bodies. Chromatin condenses and the nucleus becomes fragmented: this nuclear condensation/fragmentation, together with caspase-3 activation, represent specific and unequivocal markers of apoptosis induction.

For each sample, at least nine fields were automatically acquired, stored, and analyzed by the ArrayScan™ software. Fluorescence intensity histograms were obtained by scoring at least 800 cells with the Multiparameter Cytotoxicity 1 algorithm (Cellomics, Pittsburgh, USA), which allows acquisition of fluorescence intensity in two nuclear channels and one cytoplasmatic channel within a single measurement. Representative figures of control and treated U-2 OS cells stained with DAPI and alternatively immunostained with anti-pS6, anti-BrdU, anti-cyclin B1, anti-p-H3 and anti-active caspase3 antibodies are shown in Figure 1.

Histograms of fluorescence intensity distribution were generated using an Excel macro (Microsoft, USA) developed by the authors; the entire fluorescence range (1—4,095) of ArrayScan data was subdivided into 200 bins. Visual inspection of the fluorescence distributions related to each marker allowed determination of thresholds for the quantification of positively stained cells at each compound dose (% positive cells). This approach allowed comparison of fluorescence distributions with different dynamic ranges (i.e., for markers with different absolute fluorescence intensities), and the generation of data independently from control populations, with a consequent advantage in assay robustness. Moreover, data analysis was fast and simple, and allowed derivation of dose–response curves through which the relative compound potency toward each marker was readily apparent. The dose–response curves obtained for all compounds in the test set are reported in Figure 2.

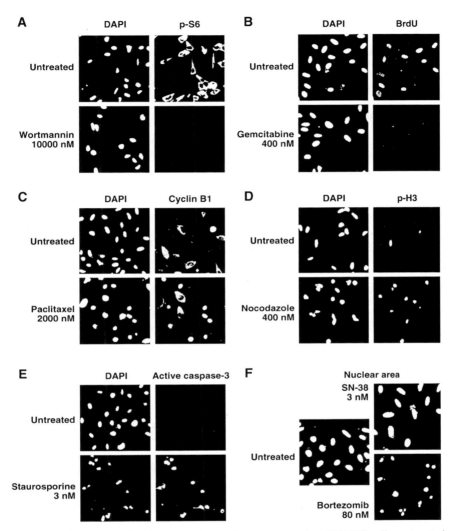

FIGURE 1. Representative images of asynchronous growing U-2 OS cells, untreated or alternatively treated for 1 h with wortmannin (**A**) and for 24 h with gemcitabine (**B**), paclitaxel (**C**), nocodazole (**D**), staurosporine (**E**), SN-38 or bortezomib (**F**) at the indicated doses. Cells were immunostained with anti-phospho-S6 (**A**), anti-BrdU (**B**), anticyclin B1 (**C**), antiphospho-H3 (**D**) or antiactive caspase-3 (**E**) antibodies, and counterstained with DAPI. Images are approximately 300 μM width.

Following compound treatment, some markers were generally found to decrease in response to drug (p-S6, cell number, BrdU incorporation), while others increased (active caspase-3) and yet others gave rise to biphasic curves, both increasing and decreasing (nuclear area, cyclin B1, p-H3). Whereas the

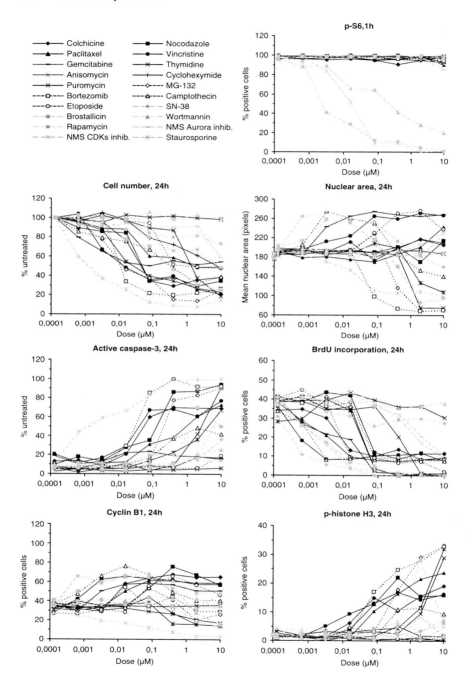

FIGURE 2. Dose–response curves. The means of at least three replicates are reported (SD not shown).

monotone curves could be analyzed by standard nonlinear regression analysis to calculate the effective concentrations 50 (EC_{50}), biphasic trends have been managed in a more complex way in order to retrieve information from both the increasing and the decreasing parts of the curve. Indeed, biphasic trends are often useful for understanding biological MOA of a compound. For example, examining the nuclear area perturbation induced by 24 h treatment with the DNA topoisomerase inhibitor SN-38 (Figures 1F and 2), it is evident that low compound doses (already at 3 nM) induce nuclear area increase, concomitantly to cell-cycle block (confirmed also by the inhibition of BrdU incorporation and the increase of cyclin B1-positive cells, Figure 2). On the contrary, higher SN-38 doses (above 0.1 µM) cause a decrease of nuclear area, due chromatin condensation subsequent to apoptosis induction (confirmed by caspase-3 activation, Figure 2). Biphasic curves generated by interpolation of experimental points were considered as the sum of two sigmoid (one increasing and one decreasing), for which the respective EC_{50} were assessed. For this reason, the profiles of cyclin B1, nuclear area and p-H3 were split into two sets (increase and decrease) for cluster analysis (Figure 3).

Hierarchical cluster analysis was performed using SpotFire software (Somerville, MA, USA) on the basis of the EC_{50} values calculated for each compound on all cellular marker. Figure 3 shows the grayscale heat-map

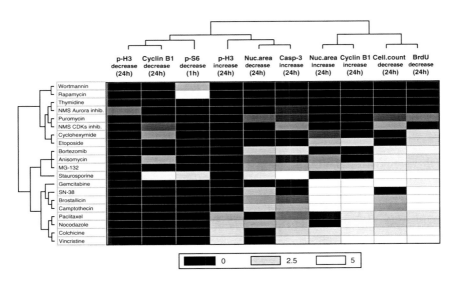

FIGURE 3. Hierarchical analysis of the 20 cell-cycle inhibitors profiles. The heat-map represents the potency of each compound towards each cellular marker (expressed as EC_{50} value). The compounds inactive at 10 µM are represented in black and the compounds with EC_{50} = 100 pM in white. Dendrograms show the reciprocal distances among compounds and markers.

representing the EC_{50} values of each compound on each cellular marker (compounds inactive at 10 µM are reported as black boxes). To construct the heat map, EC_{50} value, expressed as µM, was transformed into the logarithm of its reciprocal as follows: $a = \log10(10/EC50)$. Following this calculation, compounds inactive at 10 µM return a value of 0, whereas an EC50 of 100 pM, for example, becomes 5. This allowed us to readily emphasize slight differences among compound activities in order to aid visual comparison.

Significantly, consideration of the profile of an arbitrary set of cellular markers at different compound concentrations defines a fingerprint of the cellular and molecular phenotype associated with each compound. We observed that compounds with similar MOA tend to cluster together (Table 1 and Figure 3), confirming previous reports (Perlman et al. 2004), but in a different cellular model (U-2 OS cells) and by using a different set of cellular markers. For example, the microtubule inhibitors paclitaxel, nocodazole, colchicine, and vincristine grouped in the same cluster (Figure 3). Moreover, introduction of the variable of different treatment times enabled us, for example, to investigate the perturbations of the Akt/mTOR pathway by analyzing S6 phosphorylation after 1 h of treatment, thus avoiding any artifact or aspecific effects, such as might be caused by apoptosis, cytotoxicity, etc., arising at longer times (24 h). The validity of this strategy was confirmed by the fact that only three compounds out of 20 were found to specifically inhibit S6 phosphorylation with EC_{50} below 10 µM: the PI3K inhibitor wortmannin, the mTOR inhibitor rapamycin and the kinase inhibitor staurosporine (Figures 2–3). The phenotype observed after 24 h of treatment was dominated for the majority of the compounds by major perturbations of the cell cycle, block of proliferation and cell death induction, confirming that a longer exposure to the drugs provides information on the antiproliferative properties of drugs.

Upon inspection of biomarker clustering patterns, it becomes evident that some distinct markers give very similar responses to others over all the compounds tested. For example, nuclear area decrease is strictly associated to the degree of apoptosis as measured by caspase-3 activation; on the contrary, nuclear area increase groups with cyclin B1 expression, indicating that a G2/M block or polyploidy are associated with enlargement of nuclei. As might be expected, BrdU incorporation (a proliferation marker) and cell number decrease also belong to the same cluster.

Taken together, our data demonstrate that high-content profiling can be successfully applied to the characterization of the mechanism of action of a set of 20 cell-cycle inhibitors, and provide further validation for HTP as an effective strategy for characterizing the cellular activity and mechanism of potential new cancer drugs.

References

Arden, S.R., Janardhan, P., DiBiasio, R., Arnold, B., and Ghosh, R.N. (2002) An automated quantitative high content screening assay for neurite outgrowth. Chem. Today 20, 64–66.

Barabasz, A., Foley, B., Otto, J.C., Scott, A. and Rice, J. (2006) The use of high-content screening for the discovery and characterization of compounds that modulate mitotic index and cell cycle progression by differing mechanisms of action. Assay Drug Develop. Technol. 4(2), 153–163.

Clemons, P.A. (2004) Complex phenotypic assays in high-throughput screening. Curr. Opin. Chem. Biol. 8: 334–338.

Cohen, G.M. (1997) Caspases: the executioners of apoptosis. Biochem. J. 326, 1–16.

Ding, G.J.F., Fischer, P.A., Boltz, R.C., Schmidt, J.A., Colaianne, J.J., Gough, A., Rubin, R.A. and Miller, D.K. (1998) Characterization and quantitation of NF-kappaB nuclear translocation induced by interleukin-1 and tumor necrosis factor-alpha. Development and use of a high capacity fluorescence cytometric system. J. Biol. Chem. 273(44), 28897–28905.

Fancelli, D., Berta, D., Bindi, S., Cameron, A., Cappella, P., Carpinelli, P., Catana, C., Forte, B., Giordano, P., Giorgini, M.L., Mantegani, S., Marsiglio, A., Meroni, M., Moll, J., Pittalà, V., Roletto, F., Severino, D., Soncini, C., Storici, P., Tonani, R., Varasi, M., Vulpetti, A. and Vianello, P. (2005) Potent and selective Aurora inhibitors identified by the expansion of a novel scaffold for protein kinase inhibition. J. Med. Chem. 48(8), 3080–3084.

Fingar, D.C., Salama, S., Tsou, C., Harlow, E. and Blenis, J. (2002) Mammalian cell size is controlled by mTOR and its downstream targets S6K1 and 4EBP1/eIF4E. Genes Develop. 16(12), 1472–1487.

Gasparri, F., Cappella, P. and Galvani, A. (2006) Multiparametric cell-cycle analysis by automated microscopy. J. Biomol. Screen.

Gasparri, F., Mariani, M., Sola, F. and Galvani, A. (2004) Quantification of the proliferation index of human dermal fibroblast cultures with the ArrayScan high-content screening reader. J. Biomol. Screen. 9(3), 232–243.

Gibbs, J.B. (2000) Mechanism-based target identification and drug discovery in cancer research. Science 287, 1969–1973.

Giuliano, K.A., DeBiasio, R.L., Dunlay, R.T., Gough, A., Volosky, J.M., Zock, J., Pavlakis, G.N. and Taylor, D.L. (1997) High content screening: a new approach to easing key bottlenecks in the drug discovery process. J. Biomol. Screen. 2, 249–259.

Giuliano, K.A., Haskins, J.R. and Taylor, D.L. (2003) Advances in high content screening for drug discovery. Assay Drug Dev. Technol. 1(4), 565–577.

Gonchoroff, N.J., Katzmann, J.A., Currie, R.M., Evans, E.L., Houck, D.W., Kline, B.C., Greipp, P.R. and Loken, M.R. (1986) S-phase detection with an antibody to bromodeoxyuridine. Role of DNase pretreatment. J. Immunol. Methods 93(1), 97–101.

Hendzel, M.J., Wei, Y., Mancini, M.A., Van Hooser, A., Ranalli, T., Brinkley, B.R., Bazett-Jones, D.P. and Allis, C.D. (1997) Mitosis-specific phosphorylation of histone H3 initiates primarily within pericentromeric heterochromatin during G2 and spreads in an ordered fashion coincident with mitotic chromosome condensation. Chromosoma 106(6), 348–360.

Lang, P., Yeow, K., Nichols, A. and Scheer, A. (2006) Cellular imaging in drug discovery. Nat. Rev. Drug Discov. 5(4), 343–356.

Lövborg, H., Nygren, P. and Larsson, R. (2004) Multiparametric evaluation of apoptosis: Effects of standard cytotoxic agents and the cyanoguanidine CHS 828. Mol. Cancer Ther. 3(5), 521–526.

Minguez, J.M., Giuliano, K.A., Balachandran, R., Madiraju, C., Curran, D.P., Day, B.W. (2002) Synthesis and high content cell-based profiling of simplified analogues of the microtubule stabilizer (+)-discodermolide. Mol. Cancer Ther. 1(14), 1305–1313.

Perlman, Z.E., Mitchison, T.J. and Mayer, T.U. (2005) High-content screening and profiling of drug activity in an automated centrosome-duplication assay. Chembiochem. 6(1), 145–151.

Perlman, Z.E., Slack, M.D., Feng, Y., Mitchison, T.J., Wu, L.F. and Altschuler, S.J. (2004) Multidimensional drug profiling by automated microscopy. Science 306(5699), 1194–1198.

Pevarello, P., Brasca, M.G., Orsini, P., Traquandi, G., Longo, A., Nesi, M., Orzi, F., Piutti, C., Sansonna, P., Varasi, M., Cameron, A., Vulpetti, A., Roletto, F., Alzani, R., Ciomei, M., Albanese, C., Pastori, W., Marsiglio, A., Pesenti, E., Fiorentini, F., Bischoff, J.R. and Mercurio, C. (2005) 3-Aminopyrazole inhibitors of CDK2/cyclin A as antitumor agents. 2. Lead optimization. J. Med. Chem. 48(8), 2944–2956.

Ramm, P. (2005) Image-based screening: a technology in transition. Curr. Opin. Biotechnol. 16(1), 41–48.

Richards, G.R., Millard, R.M., Leveridge, M., Kerby, J. and Simpson, P.B. (2004) Quantitative assay of chemotaxis and chemokinesis for human neural cells. Assay Drug Dev. Technol. 2, 465–472.

Richards, G.R., Smith, A.J., Parry, F., Platts, A., Chan, G.K., Leveridge, M., Kerby, J.E. and Simpson, P.B. (2006) A morphology- and kinetics-based cascade for human neural cell high content screening. Assay Drug Dev. Tech. 4(2), 143–152.

Sherr, C.J. (1996) Cancer cell cycles. Science 274(5293), 1672–1677.

Smellie, A., Wilson, C.J. and Ng, S.C. (2006) Visualization and interpretation of high content screening data. J. Chem. Inf. Model 46(1), 201–207.

Trent, J.M., Gerner, E., Broderick, R. and Crossen, P.E. (1986) Cell cycle analysis using bromodeoxyuridine: comparison of methods for analysis of total cell transit time. Cancer Genet Cytogenet 19(1–2), 43–50.

Chapter 14
Short Abstracts – Session V

Characterization of Newly Identified Human Replication Origins

GaetanoIvanDellino

Department of Experimental Oncology, European Institute of Oncology, Milan, Italy

Replication of eukaryotic genomes is a tightly controlled process: replication initiates at several thousand origins, whose *cis*-acting sequences and *trans*-acting proteins have been best characterized in the yeast *S. cerevisiae*. Despite the early successes in the identification of microbial eukaryotic origins, the available methods for origin identification in mammalian DNA have lead to the detailed characterization of only few origins, mostly because the yeast sequence determinants (ARS consensus sequences) do not seem to be evolutionarily conserved.

We have developed a novel strategy to map human replication origins based on: (a) ultracentrifugation of fragmented cross-linked chromatin in equilibrium density gradients and (b) the finding that replication origins have a specific buoyant density. Probing tiled oligonucleotide microarrays containing the human genomic DNA from chromosome 19 with the fraction enriched for replication origins, we mapped and validated several of them. Furthermore, as it has been recently shown that post-translational chromatin modifications and nucleosome positioning can control the efficiency and/or timing of chromosomal origin activity in yeast, we started characterization of human replication origins.

Cooperative Interactions that Transform Human Cells

William C. Hahn

Department of Medical Oncology, Dana-Farber Cancer Institute, Boston, MA 02115, USA
Broad Institute of MIT and Harvard, Cambridge, MA 02142, USA

The development of cancer is a multistep process in which normal cells sustain a series of genetic alterations that together program the malignant phenotype. Much of our knowledge of cancer biology derives from the detailed study of specimens and cell lines derived from patient tumors. While these approaches continue to yield critical information regarding the identify, number, and types of alterations found in human tumors, further progress in understanding the molecular basis of malignant transformation depends upon the generation of increasingly accurate experimental models of cancer as well as more sophisticated tools to functionally annotate the cancer genome. Over the past several years, we have manipulated oncogenes, tumor suppressor genes, and telomerase to create experimental models of cancer of defined genetic constitution. Models of epithelial cancers created using these methods recapitulate many phenotypes found in patient-derived tumors and provide a foundation for the investigation of specific cancer pathways. We have used these cell systems together with activated kinase libraries to identify new oncogenes involved in the transformation of human cells. In parallel to these studies, we have developed genome scale RNAi libraries to functionally annotate the cancer genome. This lentiviral library targets each human and murine gene with five distinct constructs and permits the suppression of genes in both primary and malignant cells. Using this library, we have performed a high throughput screen to identify kinases and phosphatases that regulate mitotic progression in human cancer cells. By combining these functional approaches with information derived from mapping the structural abnormalities present in cancer genomes, we have identified several genes that contribute to cancer development. Taken together, these studies suggest that combining forward and reverse genetic approaches with information derived from the cancer genome anatomy mapping projects will yield a comprehensive list of cancer vulnerabilities.

Proteomic Strategies to the Analysis of Breast Apocrine Carcinomas

JulioE. Celis, Irina Gromova, Pavel Gromov, Josž M. A. Moreira, Teresa Cabezn, Esbern Friis, and Fritz Rank

Danish Centre for Translational Breast Cancer Research, Copenhagen, Denmark

Keywords: Apocrine carcinomas/precancerous lesions/proteomics

Breast Apocrine carcinomas represent about 0.5% of all invasive breast cancers, and despite their distinct morphological features, there are at present no standard molecular criteria available for their diagnosis. Recent proteome expression profiling studies of breast apocrine macrocysts, normal breast tissue, and breast tumors have identified specific apocrine biomarkers (15-PGDH and HMG-CoA reductase) present in early and advanced apocrine lesions. These biomarkers in combination with proteins found to be characteristically upregulated in pure apocrine carcinomas (psoriasin, S100A9, p53) provide a protein signature distinctive for benign apocrine metaplasias. These studies have also presented compelling evidence for a direct link, through the expression of 15-PGDH, between early apocrine lesions and pure apocrine carcinomas. Moreover, specific antibodies against components of the signature have identified precursor lesions in the linear progression to apocrine carcinoma. Finally, the identification of proteins that characterize the early stages of mammary apocrine differentiation such as 15-PGDH, HMG-CoA reductase, and COX-2, has opened a window of opportunity for chemoprevention.

Ultra Deep Sequencing of Amplicons Using the Genome Sequencer 20 System: The Method of Choice for Detection of Somatic Mutations

Dr. Marcus Droege

Roche Applied Science, Penzberg, Germany

The Genome Sequencer 20 System from Roche Applied Science and 454 Life Sciences, is a rapid, high throughput DNA sequencing system that routinely produces a minimum of 20 Mbp of sequence in a four and a half hour sequencing run (1). This throughput is more than 100 times faster than other conventional types of sequencing, drastically reducing time and costs. Whole bacterial genomes can be sequenced by a single individual on a single benchtop instrument in less than four days, from DNA to high coverage draft genome.

In the presentation the basic technology will be briefly described and an overview will be given on various applications performed successfully on the GS20 System up to date. The presentation will emphasize on the identification of mutations in complex cancer samples based on ultra deep sequencing of amplicons. Experimental data will provide evidence that the GS20 System is the system of choice for the very sensitive detection of rare somatic mutations without subcloning.

M. Margulies et al (2005) *Nature* 437:376–380.

Section 6
Novel Pathways and Therapeutic Targets

Chapter 15
Regulation for Nuclear Targeting of the Abl Tyrosine Kinase in Response to DNA Damage

Kiyotsugu Yoshida

Medical Research Institute, Tokyo Medical and Dental University, Tokyo 113-8510, Japan, yos.mgen@mri.tmd.ac.jp

Abstract. Abl is a ubiquitously expressed nonreceptor tyrosine kinase that is involved in diverse cellular signaling cascades. The cellular response mediated by Abl depends upon its subcellular localization. Expression of Abl in the cytoplasm results in cell proliferation and survival. In contrast, nuclear Abl is activated and induces apoptosis after genotoxic stress. Recent studies have demonstrated the molecular mechanisms by which c-Abl moves into the nucleus in the response to DNA damage. In normal cells, 14-3-3 proteins sequester c-Abl in the cytosol. Upon exposure of cells to genotoxic agents, c-jun N-terminal kinase is activated and phosphorylates 14-3-3, resulting in the release of c-Abl into the nucleus. Moreover, nuclear targeting of c-Abl is required for the induction of apoptosis in response to DNA-damaging agents. Thus, c-Abl may determine cell fate via its subcellular localization. This review summarizes the implications of these findings on our understanding of Abl-regulated cellular functions and potential therapeutic strategies to modulate the aberrant kinase.

1 Introduction

Cellular responses to DNA damage include cell-cycle arrest, activation of DNA repair, and in the event of irreparable damage, induction of apoptosis. The decision by cells to either repair DNA lesions and continue through the cell cycle or to undergo apoptosis is relevant to the incidence of mutagenesis and, subsequently, carcinogenesis. In this context, incomplete repair of DNA damage prior to replication or mitosis can result in the accumulation of heritable genetic changes. Therapeutic anticancer treatments that use DNA-damaging agents must strike a balance between induction of repair and apoptosis in order to maximize the therapeutic effect. However, the nature of the cellular signaling response that determines cell survival or cell death is far from being understood.

2 Structure and Function of Abl Tyrosine Kinase

The c-Abl tyrosine kinase is a ubiquitously expressed proto-oncogene that contains SH3, SH2, and catalytic domains in its N-terminal region (Figure 1). Contained within the C-terminus are nuclear localization motifs, a bipartite DNA-binding domain and F- and G-actin binding domains. Alternative splicing results in the expression of two c-Abl isoforms (1a and 1b), both of which are detectable in the nucleus and the cytoplasm. Recent studies revealed that c-Abl takes on an autoinhibitory conformation, and its activation requires post-translational modifications such as phosphorylation and myristoylation (Hantschel 2003; Nagar et al. 2003). Physiological functions dependent on c-Abl remain largely elusive. c-Abl is a transducer of a variety of cell extrinsic and intrinsic signals including those from growth factors, cell adhesion, oxidative stress and DNA damage (Figure 2) (Pendergast 2002). Certain insights have been derived from the findings that c-Abl shuttles between the nucleus and the cytoplasm (Taagepera 1998; Wang 2000). In contrast, oncogenic forms of Abl, including v-Abl and Bcr–Abl (Figure 1), localize exclusively in the cytoplasm and induce cellular transformation by promoting proliferation and inhibiting apoptotic cell death (Pendergast 2002; Wang 2000; Wetzler et al. 1993). In this regard, recent studies have suggested that cytoplasmic c-Abl confers cell proliferation and survival (Zhu and Wang 2004). By marked contrast, activation of nuclear c-Abl by many sources of DNA damage is associated with inhibition of cell growth and induction of apoptosis (Figure 2) (Kharbanda et al. 1998; Wang 2000). These findings

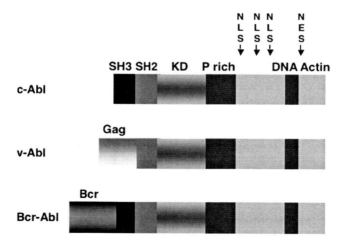

FIGURE 1. Domain structure of the Abl kinases. SH3, Src-homology-3 domain; SH2, Src homology-2 domain; KD, kinase domain; P-rich, proline-rich domain; DNA, DNA-binding domain; Actin, F-actin-binding domain; NLS, nuclear localization signal; NES, nuclear export signal.

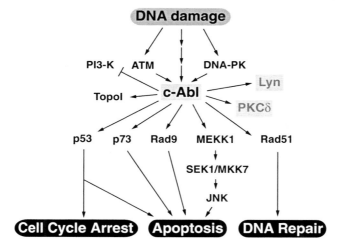

FIGURE 2. A role for c-Abl in response to DNA damage. DNA damage induces c-Abl kinase activity via ATM and other unknown mechanisms. Once activated, c-Abl can influence multiple biological functions, including cell cycle arrest, induction of apoptosis and DNA repair, by phosphorylating diverse target proteins.

indicate that the intracellular localization of c-Abl is important in dictating either survival or apoptotic responses. However, the molecular mechanism behind the nucleo-cytoplasmic shuttling of c-Abl was unknown until recently.

3 Nuclear Targeting of c-Abl in Response to DNA Damage

In our recent findings published in *Nature Cell Biology*, we demonstrated that c-Abl translocates into the nucleus in response to DNA damage or oxidative stress (Yoshida et al. 2005). The observation that the kinase-negative mutant of c-Abl also targets to the nucleus upon exposure to genotoxic stress demonstrates that the c-Abl kinase function is dispensable for nuclear translocation. Significantly, the finding that nuclear accumulation of c-Abl was relatively transient indicates that nuclear export of c-Abl is activated subsequent to the DNA damage response. The mechanism accounting for this observation is currently under investigation. Interestingly, unlike protein kinase C δ, which is also activated and moved into the nucleus after exposure to many kinds of stress (DeVries et al. 2002; Yoshida and Kufe 2001; Yoshida et al. 2002, 2003, 2006a;2006b), nuclear targeting of c-Abl was independent of its catalytic state. This suggests that other cellular proteins are responsible for the subcellular localization of c-Abl. To address this issue, we sought c-Abl-interacting proteins using liquid chromatography followed by tandem-mass spectrometry and identified several isoforms of 14-3-3 proteins.

4 Structure and Function of 14-3-3 Proteins

The 14-3-3 proteins are small (~30 kDa), acidic proteins that form both homo- and heterodimers. In human, there are seven distinct 14-3-3 genes denoted β, γ, ε, η, τ (θ), σ, and ζ while yeast and plants contain between 2 and 15 genes (Rosenquist et al. 2001; Sehnke, et al. 2002; Wang and Shakes 1996). Despite this genetic diversity, there is a surprisingly large amount of sequence identity and conservation between all the 14-3-3 isotypes (Rittinger et al. 1999; Rosenquist et al. 2000; Wang et al 1996). All the 14-3-3 proteins share a similar structure, composed of a dimerization region located at the amino-terminus and a target–protein-binding region. Crystal structure analysis and mutational studies have demonstrated that the 14-3-3 target-binding region contains residues from both the amino- and carboxy-terminal parts of the protein and revealed that 14-3-3 dimerization is mediated by a large interface composed of several regions within the amino-terminal part (Liu et al. 1995; Wang et al. 1998; Xiao et al. 1995). A number of studies indicated that the binding of 14-3-3 required target protein phosphorylation (Aitken et al. 2002). Muslin et al. used synthetic phosphopeptides based on the amino acid surrounding

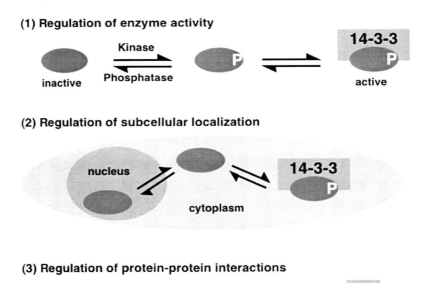

FIGURE 3. Function of 14-3-3 proteins. (1) 14-3-3 proteins regulate the activity of enzymes. (2) 14-3-3 proteins act as location anchors, controlling the subcellular localization of proteins. (3) 14-3-3 proteins can function as adaptor molecules or scaffolds resulting in the stimulation of protein–protein interactions.

Ser259 on c-Raf-1, asite of 14-3-3 binding, to define a specific sequence motif optimal for association with 14-3-3 proteins: RSXpS/TXP, where pS/T represents phospho-serine/-threonine and X any amino acid (Muslin et al. 1996).

There are a variety of role for 14-3-3 proteins reported (Figure 3). Among them, it has been well documented that they are cytoplasmic anchors that blocks import into the nucleus (Bridges and Moorhead 2005; Tzivion and Avruch 2002). The ability of 14-3-3 proteins to potentiate the cytoplasmic localization of binding partners has dramatic effects on signal transduction cascades, cell cycle progression, regulation of apoptotic pathways or cytoskeletal organization (Aitken et al. 2002; Bridges et al. 2005; Tzivion et al. 2002).

5 14-3-3 Sequesters c-Abl in the Cytoplasm

We found that c-Abl is tethered in the cytoplasm by binding to 14-3-3 proteins and that the c-Abl-14-3-3 complexes are disrupted after genotoxic stress. Functional binding analyses demonstrated that this binding depends on phosphorylation of c-Abl on Thr735, which is included within the 14-3-3 binding motif RSXpS/TXP. Importantly, Thr735 is located between the second and third nuclear localization signals (NLS) in the C-terminus of c-Abl. We therefore hypothesized that binding of 14-3-3 to phosphorylated Thr735 masks the c-Abl NLS and prevent its nuclear entry. This is similar to the mode of regulation reported for cdc25c, FKHRL1, DAF-16, PKU and histone deacetylase (Muslin and Xing 2000).

Previous studies have shown that the phosphorylation level of the Ser/Thr residues within the 14-3-3 binding motif determines the binding affinity between 14-3-3 and target proteins (Muslin et al 2000). To assess whether this is the case with c-Abl, we monitored the phosphorylation status of Thr735 following treatment of cells with anti-cancer agents. Unexpectedly, the phosphorylation status of Thr735 was unaffected by DNA damage, indicating that the disruption of the c-Abl-14-3-3 complex is independent of Thr735 phosphorylation. These findings led us to examine the alternative possibility that modification of 14-3-3 proteins following genotoxic stress regulates the binding to c-Abl. Of note, Thr735 kinase is currently unknown. It is possible that the expression or activation of this unknown kinase(s) may be regulated in a tissue- or developmental-specific manner, or may be altered under abnormal growth conditions, thereby providing another level for the regulation of nuclear targeting of c-Abl and c-Abl-dependent apoptosis.

6 JNK Phosphorylation of 14-3-3 Induces Dissociation of c-Abl

Recent studies have shown that JNK phosphorylation of 14-3-3 is associated with its dissociation from Bax or Bad (Sunayama et al. 2005; Tsuruta et al. 2004). Cellular stresses induced JNK-mediated 14-3-3ζ phosphorylation at

Ser184 (Tsuruta et al. 2004). Importantly, phosphorylation of 14-3-3 by JNK caused the dissociation of Bax from 14-3-3, leading to Bax translocation to the mitochondria and to apoptosis (Tsuruta et al. 2004). In addition, JNK phosphorylation of 14-3-3 also released the proapoptotic proteins Bad and FOXO3a from 14-3-3 and antagonized the effects of Akt signaling (Sunayama et al. 2005). Other studies have demonstrated that cleavage of 14-3-3 by caspase-3 promotes cell death releasing Bax or Bad from 14-3-3 and facilitates their translocation to the mitochondria and their interaction with Bcl-2 or Bcl-xL (Nomura et al. 2003; Won et al. 2003). We examined both possibilities to elucidate the mechanism by which 14-3-3 releases c-Abl upon exposure of cells to genotoxic stress. The results demonstrated that release of c-Abl from 14-3-3 depends upon phosphorylation of 14-3-3 proteins by the c-Jun N-terminal kinase (JNK), a kinase activated in response to DNA damage or oxidative stress. It remains to be elucidated how JNK phosphorylation modulates 14-3-3 dissociation from target proteins. However, the presence of JNK phosphorylation consensus sites within the 14-3-3 target-binding domain suggests that phosphorylation by JNK could impair the association of proteins with the 14-3-3 binding domain. Alternatively, JNK phosphorylation might induce conformational changes within 14-3-3 proteins, thereby causing the dissociation of target proteins.

7 Nuclear Targeting of c-Abl is Essential for Induction of Apoptosis

To assess the physiological relevance of our findings, we examined the role of 14-3-3 in c-Abl-mediated apoptosis. The data demonstrated that 14-3-3 attenuated c-Abl-induced apoptosis and that, following DNA damage, a 14-3-3 mutant resistant to JNK phosphorylation conferred a more protective effect on c-Abl-mediated apoptosis than wild-type 14-3-3. Thus, these findings collectively support a model in which JNK phosphorylates 14-3-3 proteins in the cytoplasm to cause release of c-Abl, resulting in nuclear accumulation of c-Abl and induction of apoptosis (Figure 4). We previously demonstrated that nuclear c-Abl induces apoptosis in response to DNA damage (Huang, et al.

FIGURE 4. A model for nuclear targeting of c-Abl in response to DNA damage. (A) Under normal conditions, c-Abl undergoes nucleo-cytoplasmic shuttling. Phosphorylated c-Abl binds to 14-3-3 proteins, which sequesters c-Abl in the cytoplasm. (B) Upon DNA damage, cytoplasmic sequestration of c-Abl is abrogated by JNK-mediated 14-3-3 phosphorylation. Dissociated c-Abl is targeted to the nucleus, nuclear c-Abl is activated, and apoptosis is induced.

15. Regulation for Nuclear Targeting of the Abl Tyrosine Kinase   161

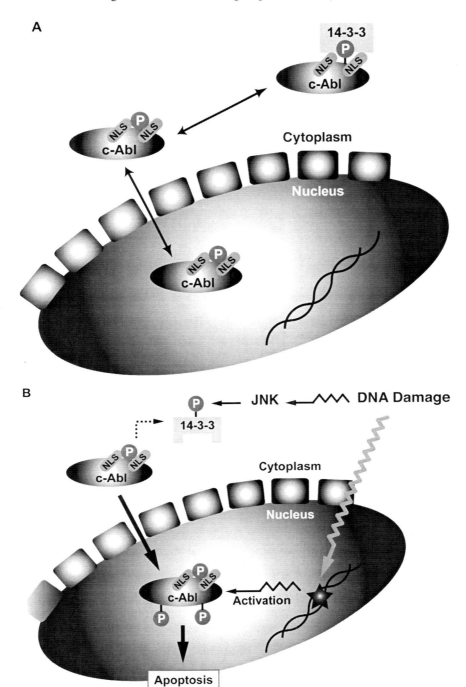

1997; Yuan et al. 1997). Subsequently, we identified that activation of nuclear proteins, such as p73 and Rad9, by nuclear c-Abl phosphorylation is associated with DNA damage-induced cell death (Yoshida et al. 2002; Yuan et al. 1999). Together with these results, our recent findings further support an essential role for c-Abl in the determination of cell fate by its nuclear shuttling in response to DNA damage (Yoshida et al. 2005).

8 Potential Therapeutic Strategies Based on Regulating Nuclear Translocation of Abl Kinases

The oncogenic form of Abl kinase, Bcr–Abl, is constitutively active and localizes to the cytoplasm. Bcr–Abl is the underlying cause in the development and progression of chronic myeloid leukemia (CML). It remains unknown whether 14-3-3 proteins sequester Bcr–Abl in the cytoplasm. However, like c-Abl, Bcr–Abl contains three NLSs, indicating the possibility that 14-3-3 and JNK might regulate subcellular localization of Bcr–Abl. In this context, a recent study suggested that Bcr–Abl translocates from the cytoplasm to the nucleus following treatment of cells with the topoisomerase II inhibitor etoposide (Dierov et al. 2004). Moreover, it is intriguing that several 14-3-3 family proteins also interact with the Bcr region of Bcr–Abl (Reuther et al. 1994). Thus, these findings collectively support the possibility that release of Bcr–Abl from 14-3-3 by JNK phosphorylation following DNA damage contributes to the nuclear accumulation of Bcr–Abl. Another report documented that Bcr–Abl is found in the nucleus after treatment of cells with the Abl kinase inhibitor STI571 and the nuclear export inhibitor leptomycin B (Vigneri and Wang 2001). Importantly, accumulation of Bcr–Abl in the nucleus is associated with induction of apoptosis. This result is consistent with our finding that activation of nuclear Abl kinase is required for DNA damage-induced apoptosis (Yoshida et al. 2005). In this regard, trapping Bcr–Abl in the nucleus is an interesting approach to treat CML without severe side effects. Based on our findings, there are several potential therapies that can be suggested. For example, inhibiting the binding of c-Abl/Bcr–Abl to 14-3-3 proteins by small compounds might effectively cause nuclear accumulation of c-Abl/Bcr–Abl and induce apoptosis. Dephosphorylation of Thr735 on c-Abl/Bcr–Abl could be another viable approach. Whether 14-3-3 and JNK regulate nuclear targeting of Bcr–Abl following DNA damage is still a question; however, these therapeutic possibilities are worth examining in future work and open up a new avenue of research into molecular-targeted cancer therapies.

Acknowledgments. This work was supported by grants from the Ministry of Education, Science and Culture of Japan, Kanae Foundation for Life & Socio-medical Science and Kowa Life Science Foundation.

References

Aitken, A., Baxter, H., Dubois, T., Clokie, S., Mackie, S., Mitchell, K., Peden, A. and Zemlickova, E. (2002) Specificity of 14-3-3 isoform dimer interactions and phosphorylation. Biochem Soc Trans 30, 351–360.

Bridges, D. and Moorhead, G. B. (2005) 14-3-3 proteins: a number of functions for a numbered protein. Sci STKE 2005, re10.

DeVries, T. A., Neville, M. C. and Reyland, M. E. (2002) Nuclear import of PKCdelta is required for apoptosis: identification of a novel nuclear import sequence. EMBO J 21, 6050–6060.

Dierov, J., Dierova, R. and Carroll, M. (2004) BCR/ABL translocates to the nucleus and disrupts an ATR-dependent intra-S phase checkpoint. Cancer Cell 5, 275–285.

Hantschel, O., Nagar, B., Guettler, S., Kretzschmar, J., Dorey, K., Kuriyan, J. and Superti-Furga, G. (2003) A myristoyl/phosphotyrosine switch regulates c-Abl. Cell 112, 845–857.

Huang, Y., Yuan, Z. M., Ishiko, T., Nakada, S., Utsugisawa, T., Kato, T., Kharbanda, S. and Kufe, D. W. (1997) Pro-apoptotic effect of the c-Abl tyrosine kinase in the cellular response to 1-β-D-arabinofuranosylcytosine. Oncogene 15, 1947–1952.

Kharbanda, S., Yuan, Z. M., Weichselbaum, R. and Kufe, D. (1998) Determination of cell fate by c-Abl activation in the response to DNA damage. Oncogene 17, 3309–3318.

Liu, D., Bienkowska, J., Petosa, C., Collier, R. J., Fu, H. and Liddington, R. (1995) Crystal structure of the ζ isoform of the 14-3-3 protein. Nature 376, 191–194.

Muslin, A. J., Tanner, J. W., Allen, P. M. and Shaw, A. S. (1996) Interaction of 14-3-3 with signaling proteins is mediated by the recognition of phosphoserine. Cell 84, 889–897.

Muslin, A. J. and Xing, H. (2000) 14-3-3 proteins: regulation of subcellular localization by molecular interference. Cell Signal 12, 703–709.

Nagar, B., Hantschel, O., Young, M. A., Scheffzek, K., Veach, D., Bornmann, W., Clarkson, B., Superti-Furga, G. and Kuriyan, J. (2003) Structural basis for the autoinhibition of c-Abl tyrosine kinase. Cell 112, 859–871.

Nomura, M., Shimizu, S., Sugiyama, T., Narita, M., Ito, T., Matsuda, H. and Tsujimoto, Y. (2003) 14-3-3 Interacts directly with and negatively regulates pro-apoptotic Bax. J Biol Chem 278, 2058–2065.

Pendergast, A. M. (2002) The Abl family kinases: mechanisms of regulation and signaling. Adv Cancer Res 85, 51–100.

Reuther, G. W., Fu, H., Cripe, L. D., Collier, R. J. and Pendergast, A. M. (1994) Association of the protein kinases c-Bcr and Bcr–Abl with proteins of the 14-3-3 family. Science 266, 129–133.

Rittinger, K., Budman, J., Xu, J., Volinia, S., Cantley, L. C., Smerdon, S. J., Gamblin, S. J. and Yaffe, M. B. (1999) Structural analysis of 14-3-3 phosphopeptide complexes identifies a dual role for the nuclear export signal of 14-3-3 in ligand binding. Mol Cell 4, 153–166.

Rosenquist, M., Alsterfjord, M., Larsson, C. and Sommarin, M. (2001) Data mining the Arabidopsis genome reveals fifteen 14-3-3 genes. Expression is demonstrated for two out of five novel genes. Plant Physiol 127, 142–149.

Rosenquist, M., Sehnke, P., Ferl, R. J., Sommarin, M. and Larsson, C. (2000) Evolution of the 14-3-3 protein family: does the large number of isoforms in multicellular organisms reflect functional specificity? J Mol Evol 51, 446–458.

Sehnke, P. C., Rosenquist, M., Alsterfjord, M., DeLille, J., Sommarin, M., Larsson, C. and Ferl, R. J. (2002) Evolution and isoform specificity of plant 14-3-3 proteins. Plant Mol Biol 50, 1011–1018.

Sunayama, J., Tsuruta, F., Masuyama, N. and Gotoh, Y. (2005) JNK antagonizes Akt-mediated survival signals by phosphorylating 14-3-3. J Cell Biol 170, 295–304.

Taagepera, S., McDonald, D., Loeb, J. E., Whitaker, L. L., McElroy, A. K., Wang, J. Y. and Hope, T. J. (1998) Nuclear-cytoplasmic shuttling of C-ABL tyrosine kinase. Proc Natl Acad Sci U S A 95, 7457–7462.

Tsuruta, F., Sunayama, J., Mori, Y., Hattori, S., Shimizu, S., Tsujimoto, Y., Yoshioka, K., Masuyama, N. and Gotoh, Y. (2004) JNK promotes Bax translocation to mitochondria through phosphorylation of 14-3-3 proteins. EMBO J 23, 1889–1899.

Tzivion, G. and Avruch, J. (2002) 14-3-3 proteins: active cofactors in cellular regulation by serine/threonine phosphorylation. J Biol Chem 277, 3061–3064.

Vigneri, P. and Wang, J. Y. (2001) Induction of apoptosis in chronic myelogenous leukemia cells through nuclear entrapment of BCR–ABL tyrosine kinase. Nat Med 7, 228–234.

Wang, H., Zhang, L., Liddington, R. and Fu, H. (1998) Mutations in the hydrophobic surface of an amphipathic groove of 14-3-3ζ disrupt its interaction with Raf-1 kinase. J Biol Chem 273, 16297–16304.

Wang, J. Y. (2000) Regulation of cell death by the Abl tyrosine kinase. Oncogene 19, 5643–5650.

Wang, W. and Shakes, D. C. (1996) Molecular evolution of the 14-3-3 protein family. J Mol Evol 43, 384–398.

Wetzler, M., Talpaz, M., Van Etten, R. A., Hirsh-Ginsberg, C., Beran, M. and Kurzrock, R. (1993) Subcellular localization of Bcr, Abl, and Bcr–Abl proteins in normal and leukemic cells and correlation of expression with myeloid differentiation. J Clin Invest 92, 1925–1939.

Won, J., Kim, D. Y., La, M., Kim, D., Meadows, G. G. and Joe, C. O. (2003) Cleavage of 14-3-3 protein by caspase-3 facilitates bad interaction with Bcl-x_L during apoptosis. J Biol Chem 278, 19347–19351.

Xiao, B., Smerdon, S. J., Jones, D. H., Dodson, G. G., Soneji, Y., Aitken, A. and Gamblin, S. J. (1995) Structure of a 14-3-3 protein and implications for coordination of multiple signalling pathways. Nature 376, 188–191.

Yoshida, K., Komatsu, K., Wang, H. G. and Kufe, D. (2002) c-Abl tyrosine kinase regulates the human Rad9 checkpoint protein in response to DNA damage. Mol Cell Biol 22, 3292–3300.

Yoshida, K. and Kufe, D. (2001) Negative regulation of the SHPTP1 protein tyrosine phosphatase by protein kinase C δ in response to DNA damage. Mol Pharmacol 60, 1431–1438.

Yoshida, K., Liu, H. and Miki, Y. (2006) Protein Kinase C δ regulates Ser46 phosphorylation of p53 tumor suppressor in the apoptotic response to DNA damage. J Biol Chem 281, 5734–5740.

Yoshida, K., Miki, Y. and Kufe, D. (2002) Activation of SAPK/JNK signaling by protein kinase Cδ in response to DNA damage. J Biol Chem 277, 48372–48378.

Yoshida, K., Wang, H. G., Miki, Y. and Kufe, D. (2003) Protein kinase Cδ is responsible for constitutive and DNA damage-induced phosphorylation of Rad9. EMBO J 22, 1431–1441.

Yoshida, K., Yamaguchi, T., Natsume, T., Kufe, D. and Miki, Y. (2005) JNK phosphorylation of 14-3-3 proteins regulates nuclear targeting of c-Abl in the apoptotic response to DNA damage. Nat Cell Biol 7, 278–285.

Yoshida, K., Yamaguchi, T., Shinagawa, H., Taira, N., Nakayama, K. I. and Miki, Y. (2006) Protein Kinase C δ Activates Topoisomerase IIα To Induce Apoptotic Cell Death in Response to DNA Damage. Mol Cell Biol 26, 3414–3431.

Yuan, Z. M., Huang, Y., Ishiko, T., Kharbanda, S., Weichselbaum, R. and Kufe, D. (1997) Regulation of DNA damage-induced apoptosis by the c-Abl tyrosine kinase. Proc Natl Acad Sci U S A 94, 1437–1440.

Yuan, Z. M., Shioya, H., Ishiko, T., Sun, X., Gu, J., Huang, Y. Y., Lu, H., Kharbanda, S., Weichselbaum, R. and Kufe, D. (1999) p73 is regulated by tyrosine kinase c-Abl in the apoptotic response to DNA damage. Nature 399, 814–817.

Zhu, J. and Wang, J. Y. (2004) Death by Abl: a matter of location. Curr Top Dev Biol 59, 165–192.

Chapter 16
Short Abstracts – Session VI

Cooperative Interactions that Transform Human Cells

William C. Hahn

Department of Medical Oncology, Dana-Farber Cancer Institute, Boston, MA 02115 - Broad Institute of MIT and Harvard, Cambridge, MA 02142, USA

The development of cancer is a multistep process in which normal cells sustain a series of genetic alterations that together program the malignant phenotype. Much of our knowledge of cancer biology derives from the detailed study of specimens and cell lines derived from patient tumors. While these approaches continue to yield critical information regarding the identify, number, and types of alterations found in human tumors, further progress in understanding the molecular basis of malignant transformation depends upon the generation of increasingly accurate experimental models of cancer as well as more sophisticated tools to functionally annotate the cancer genome. Over the past several years, we have manipulated oncogenes, tumor suppressor genes and telomerase to create experimental models of cancer of defined genetic constitution. Models of epithelial cancers created using these methods recapitulate many phenotypes found in patient-derived tumors and provide a foundation for the investigation of specific cancer pathways. We have used these cell systems together with activated kinase libraries to identify new oncogenes involved in the transformation of human cells. In parallel to these studies, we have developed genome scale RNAi libraries to functionally annotate the cancer genome. This lentiviral library targets each human and murine gene with five distinct constructs and permits the suppression of genes in both primary and malignant cells. Using this library, we have performed a high throughput screen to identify kinases and phosphatases that regulate mitotic progression in human cancer cells. By combining these functional approaches with information derived from mapping the structural abnormalities present in cancer genomes, we have identified several genes that contribute to cancer development. Taken together, these studies suggest that combining forward and reverse genetic approaches with information derived from the cancer genome anatomy mapping projects will yield a comprehensive list of cancer vulnerabilities.

Proteomic Strategies to the Analysis of Breast Apocrine Carcinomas

Julio E. Celis, Irina Gromova, Pavel Gromov, José M. A. Moreira, Teresa Cabez, Esbern Friis Hansen, and Fritz Rank

Danish Centre for Translational Breast Cancer Research, Copenhagen, Denmark

Keywords: Apocrine carcinomas/precancerous lesions/proteomics

Breast Apocrine carcinomas represent about 0.5% of all invasive breast cancers, and despite their distinct morphological features, there are at present no standard molecular criteria available for their diagnosis. Recent proteome expression profiling studies of breast apocrine macrocysts, normal breast tissue, and breast tumors have identified specific apocrine biomarkers (15-PGDH and HMG-CoA reductase) present in early and advanced apocrine lesions. These biomarkers in combination with proteins found to be characteristically upregulated in pure apocrine carcinomas (psoriasin, S100A9, p53) provide a protein signature distinctive for benign apocrine metaplasias. These studies have also presented compelling evidence for a direct link, through the expression of 15-PGDH, between early apocrine lesions and pure apocrine carcinomas. Moreover, specific antibodies against components of the signature have identified precursor lesions in the linear progression to apocrine carcinoma. Finally, the identification of proteins that characterize the early stages of mammary apocrine differentiation such as 15-PGDH, HMG-CoA reductase, and COX-2, has opened a window of opportunity for chemoprevention.

Ultra Deep Sequencing of Amplicons Using the Genome Sequencer 20 System

Dr. Marcus Droege

Roche Applied Science, Penzberg, Germany

The Genome Sequencer 20 System from Roche Applied Science and 454 Life Sciences, is a rapid, high throughput DNA sequencing system that routinely produces a minimum of 20 Mbp of sequence in a four and a half hour sequencing run (1). This throughput is more than 100 times faster than other conventional types of sequencing, drastically reducing time and costs. Whole bacterial genomes can be sequenced by a single individual on a single benchtop instrument in less than four days, from DNA to high coverage draft genome.

In the presentation the basic technology will be briefly described and an overview will be given on various applications performed successfully on the GS20 System up to date. The presentation will emphasize on the identification of mutations in complex cancer samples based on ultra deep sequencing of amplicons. Experimental data will provide evidence that the GS20 System is the system of choice for the very sensitive detection of rare somatic mutations without subcloning.

M. Margulies et al (2005) *Nature* 437:376–380.

High Throughput 3D Correlative Microscopy**

CarloTacchetti

IFOM Centre of Cell Oncology and Ultrastructure, Department of Experimental Medicine, University of Genoa, Genova Italy

Cortese K.*∞, Canfora M.*∞, Vicidomini G.*∞, Gagliani MC*∞, Mattioli L.*∞, Boccacci P.*∞, Diaspro A.*∞, Tacchetti C.∞**MicroscoBio Research Center, University of Genova, and ∞IFOM, Milano, Italy**

By correlative microscopy, the same structures observed at the fluorescence light microscopy (FLM) can be analyzed at the EM level. Correlative microscopy resolves several different limitations of fluorescence microscopy. In particular, the absence of a "reference space," e.g., the unlabeled structures not seen in FLM which do not contribute to the analysis. Therefore, it is possible to answer to question as: what is the true size and shape of the structure observed by FLM? However, these studies are extremely time-consuming, especially if statistically relevant data have to be acquired. To overcome these limitations we have set a new correlative method, based on the use semi-ultrathin cryo-sections, designed to: (a) Increase to hundreds the number of events that can be correlated, at each single microscopy session. (b) Obtain 3D correlation between FLM and EM. (c) Provide a free-share software, designed on ImageJ, to perform, in semi- or fully- automatic way, most of the operations needed for the multi section 3D reconstruction of FLM images. (d) Improve the contrast along the Z-axis, at FLM observation, due to the utilization of sections with a subresolution thickness. (e) Allow for further detection of different antigens on the correlated structures at the EM level.

Section 7
Poster Abstracts

Ubiquitination Integrates EGFR Signaling and Attenuation

P.P. DiFiore

IFOM, The FIRC Institute for Molecular Oncology, 20134 Milan, Italy

Work in many laboratories in the past few years has established that monoubiquitination (or oligo-ubiquitination) can function as a signaling device to establish protein:protein interactions with intracellular proteins (Ub receptors) harboring ubiquitin-recognition modules. The list of such modules includes now several domains and devices, such as UIM, CUE, UEV, UBA and others. Thus, there are important similarities between the network of Ub-mediated interactions and other extensively characterized networks similarly based on post-translational modifications, such as tyrosine phosphorylation. Our lab is interested in investigating how ubiquitination couples RTKs (receptor tyrosine kinases) with downstream pathways. We have shown previously that some RTKs, such as the EGFR, long thought to be polyubiquitinated, are actually monoubiquitinated. We will now present data showing that ubiquitination of the EGFR allows its targeting to a nonclathrin pathway of internalization, through its binding to UIM-containing proteins, such as eps15 and epsin. Such a nonclathrin pathway might represent the preferred pathway for receptor downregulation and degradation, while the clathrin pathway might be preferentially associated with receptor recycling to the cell surface and sustained effector signaling.

Human Tumor Antigens, Successful Immuno Surveillance and Prophylactic Cancer Vaccines

Olivera J. Finn, Ph.D.

Department of Immunology, University of Pittsburgh School of Medicine, Pittsburgh, Pennsylvania, USA

Numerous shared human tumor antigens have been characterized and a large number of them have been shown effective in preventing tumor growth in animal models. Based on in vitro data and animal models, some have been tested as components of cancer vaccines in Phase I and II clinical trials. We have previously reported on two shared human tumor antigens, MUC1 and cyclin B1 (CB1), and on specific differences between their expressions on normal cells versus tumor cells. We have shown that these differences are key to their immunogenicity and that in transgenic animal models, immune responses induced to their tumor forms protect from tumor challenge. MUC1 has also been tested in several Phase I/II trials by us and other investigators around the world. In order to pave the way for use of MUC1 and CB1 cancer vaccines for cancer prevention, we have focused on obtaining information and will report on the following topics: (1) Expression of these antigens on cancer stem cells; (2) Expression of these antigens on premalignant lesions; (3) Evidence for successful immunosurveillance of these antigens in spontaneous mouse tumor models; and (4) Evidence for successful immunosurveillance of these antigens in humans.

Functional Analysis of the BRCA Gene Products

David M. Livingston

Department of Genetics, Dana Farber Cancer Institute, Boston, USA

BRCA1 and *2* are genes that, when inherited in a mutant form, predispose to breast and ovarian cancer. The products of these genes are large, multifunctional proteins that concentrate in the cell nucleus and prepare cells to respond to genomic insults. How in molecular terms these molecules perform these complex and life preserving functions and how these operations are translated into breast and ovarian cancer suppression are fundamentally important questions in the field. Efforts to address these questions and the results thereof will be described in my lecture.

Transcriptional Targets of Acute Myeloid Leukemia Fusion Proteins: A High-Throughput Approach

Alessandro Gardini and Myriam Alcalay

European Institute of Oncology and IFOM Foundation, Milan, Italy

The pathogenesis of acute myeloid leukemias (AMLs) is linked to the function of oncogenic fusion proteins generated as a consequence of chromosome translocations. AML-associated fusion proteins function as aberrant transcriptional regulators that interfere with the process of myeloid differentiation, determine a stage-specific arrest of maturation and enhance survival in a cell-type specific manner. The abnormal regulation of transcriptional networks occurs through mechanisms that include recruitment of aberrant co-repressor complexes, alterations in chromatin remodeling, and disruption of specific subnuclear compartments. We integrated high-throughput approaches to investigate transcriptional regulation in AML. First, we analyzed gene expression profiles of cells expressing two AML fusion proteins (AML1/ETO, PML/RARα) and identified a set of putative target genes. Next, we performed genome-wide chromatin immunoprecipitation assays to look for direct target genes and specific histone modifications. By merging the data sets, we identified genes whose expression is deregulated by direct binding of fusion protein to regulatory regions. Two surprising observations emerged: (i) genes that are part of the same functional pathways are co-regulated; (ii) in contrast to what is expected from a transcriptional repressor a relevant proportion of genes are induced by AML fusion proteins, suggesting that AML fusion proteins can also act as transcriptional activators.

Sumoylation and Ubiquitination Regulate the PEA3 Group Member Transcription-Enhancing Activity

Yvan de Launoit[1,2], Jean-Luc Baert[1], Cindy degerny[1], Claude Beaudoin[1,2], Didier Monte[1], and Sebastien Mauen[1,2]

UMR 8161, CNRS / Université de Lille 1 / Université de Lille 2 / Institut Pasteur de Lille, Institut de Biologie de Lille, BP 447, 1 rue Calmette, 59021 Lille Cedex, France. 2 Laboratoire de Virologie Moléculaire, Faculté de Medecine, Universite Libre de Bruxelles, CP 614, 808 route de Lennik, 1070 Brussels, Belgium

The PEA3 group of Ets transcription factors is composed of three highly conserved members: Erm, Er81 and Pea3. Among their target genes are several matrix metalloproteases which are enzymes degrading extracellular matrix during cancerous metastasis process. In fact, PEA3 group members are often overexpressed in different types of cancers which also overexpress these MMP and present a disseminating phenotype. They regulate transcription of their target genes following post-translational modifications, such as phosphorylation. We then investigated whether modifications on lysine residues could affect the transcriptional activity of Erm, a prototype of the PEA3 group member. After showing that Erm degradation depends on the 26S proteasome pathway, we propose that ubiquitination enhances the transcriptional activity of the protein. We also identified that Erm is sumoylated at positions 89, 263, 293 and 350, a modification which causes inhibition of ERM-dependent transcription without affecting the protein's subcellular localization, stability, or DNA-binding. We will discuss on the putative cross-talk of these post-translational modifications on the activity of the PEA3 group members.

Mechanism of P53/P66 Pathway in Regulation of Apoptosis ROS and Aging

E. Migliaccio, D. Mascheroni, M. Cesaroni, L. Luzi, M. Giorgio, and PG. Pelicci

European Institute of Oncology, Firc Institute of Molecular Oncology, Milan, Italy

Biochemical and biological evidences showed that p66shc protein is a downstream target of p53 tumor suppression protein. *p66Shc* was the first mammalian gene whose mutation was demonstrated to increase resistance to oxidative stress and to prolong life span. Most important, we have established a new and surprising redox property of p66Shc protein. We characterized a p66Shc/cytochrome c electron transfer reaction that might give an alternative mechanism of generating H_2O_2 in mitochondria. We found that p66shc is indispensable for the ability of activated p53 to induce elevation of ROS and apoptosis while other functions of p53 are not influenced. Notably, p66Shc-/- mice that have an increase of 30% of lifespan have a reduced risk of tumor, while transgenic animals with constitutive expression of activated p53 show decreased cancer incidence and accelerated aging. However, since aging and cancer are biological linked study of p53–p66Shc signaling pathway could be one of the sources to investigate on cancer. We studied the involvement of the p66 redox balance on transcription by microarray experiments. Preliminary experiments suggest that p53–p66 pathway regulate of a set of gene that control apoptosis and mitosis transition. The relationship between p53–p66Shc-mediated apoptosis and organism aging is discussed.

Met Addiction Sustains Invasive Growth

Simona Corso, Cristina Migliore; Giovanni De Rosa, Paolo M Comoglio, and Silvia Giordano

University of Torino, Medical School, Institute for Cancer Research and Treatment

A number of oncogenes have been implicated in the development of human tumors, but it is still unclear if their activation is mandatory during all the phases of tumor progression toward invasion and metastasis. The Met proto-oncogene, encoding the tyrosine kinase receptor for hepatocyte growth factor (HGF), controls "invasive growth", a physiological genetic program leading to cell growth, invasion and protection from apoptosis. Its role in human tumors and in metastasis development has been well documented. We show that silencing the expression of the endogenous Met proto-oncogene in tumor cells by an inducible lentiviral delivery system of RNA interference impaired the ability of tumor cells to execute the full invasive growth program in vitro and to generate tumors and metastases in vivo. Moreover, silencing Met expression after metastasis formation resulted in their regression. These experiments indicate that "addiction" to Met is a promising therapeutic target, even in the advanced phases of tumor development.

Preleukemic Phase Analysis to Understand the Biological Contribution of PML-RARα to the Leukemogenic Process

Valeria Cambiaghi, Piergiuseppe Pelicci

European Institute of Oncology, Via Ripamonti 435, Milan, Italy

Acute myeloid leukemia (AML) associated fusion proteins induce a preleukemic state. The t(15;17) is found exclusively in acute promyelocytic leukaemia (APL), a subtype of AML, and codes for the PML-RARα fusion protein which incorporate the DNA-binding domain of the RARα transcription factor and that functions as a constitutive repressor of retinoic acid (RA)-target genes. In transgenic mice expressing PML-RARα in the bone marrow, preleukemia occurs after accumulation of additional genetic lesions. In this case preleukemic bone marrow appears morphologically normal before the second hit event. The biological mechanisms underlying the contribution of fusion proteins to the preleukemic phase and to the maintenance of the transformed leukemic phenotype are unknown. Preliminary data demonstrate how PML-RARα in vivo drives the expansion of the stem cell compartment: in fact, preleukemic bone marrow is characterized by an overproliferation in Lin- c-Kit+ Sca-1+ subpopulation associated with an overexpression of stem cell genes. Moreover, in bone marrow competition experiments, preleukemic cells have a proliferative advantage over normal cells. Now we are investigating which is the mechanism used by PML-RARα to induce genome instability: the answer seems to be a slow mechanism of telomere shortening during the preleukemic phase.

A Novel Enhancer of the PDGF Beta-Receptor Gene is Activated by GATA, Inducing a Differentiation of Neuroblastoma

Weiwen Yang, Masaharu Kaneko[1,2], Yoshiki Matsumoto[1], Fujiko Watt[1,3], and Keiko Funa[1]

[1] Department of Cell Biology, Institute of Anatomy and Cell Biology, G™teborg University, Box 420, SE-405 30 Gothenburg, Sweden
[2] Present address, Department of Neurosurgery, Shiga University of Medical Science, Seta, Tsukinowa-cho, Otsu, Shiga, 520-2192, Japan
[3] Present address, Children's Cancer Institute Australia for Medicine, Australia

PDGF acts as an autocrine factor in certain tumors in which the PDGF beta-receptor expression is upregulated. In order to elucidate the control mechanism for the receptor expression, we have isolated an enhancer from two P1 clones that together contain 102 Kb NotI region covering the entire human *PDGFRB* gene. They were cloned into the PDGFRB-enhancer trap vector to make a library for identification of enhancer DNA. One of the best enhancer clones was identified and further examined by making several deletion mutants in a luciferase vector. This core enhancer was most active in neuroblastoma cell lines, IMR32 and BE2, but less active in a hemangioma as well as in a smooth muscle cell line. Chip assay revealed that SP1, AP2, and GATA2, were shown to bind the core enhancer in BE2 cells. Their interaction occurred dependently of the cell cycle and also synchronously with their binding to the promoter. Transfection of GATA2 alone or with Ets, binding adjacent to GATA, resulted in differentiation of BE2 cells in parallel with increased expression of PDGF beta-receptor. Thus, the kinetic of the enhancer activation honestly reflect the receptor expression during cell cycle and upon neuronal differentiation.

Development of Retroviral Microarrays for New Drug Target Identification and Gene Therapy Application

R. Carbone[1], I. Marangi[1], A. Zanardi[1], L. Giorgetti[2], E. Clerici[2,3], G. Bongiorno[2], P. Piseri[2], P. Milani[2], P.G. Pelicci[1]

[1] Dipartimento di Oncologia Sperimentale, Istituto Europeo di Oncologia 435, Via Ripamonti, 20141 Milano, Italy
[2] (CIMAINA) Centro Interdisciplinare Materiali e Interfacce Nanostrutturati, Dipartimento di Fisica, Universita' di Milano, Italy
[3] Tethis S.r.l. Piazzetta Bossi, 4 20121 Milano, Italy

Microarrays have become a key technology of the post-genomic era of cell biology. New powerful tools for functional genomics have been recently described (1, 2) which consists of cell-based microarray for high-throughput analysis of gene function. A major limitation is that only few cell lines (293T, COS, Hela) can be used as cellular target (3, 4). We have developed a new cell-based microarray for phenotype screening on primary cells by means of retrovirus immobilization on slide. Recent work (5) demonstrated the efficiency of biotinylated retrovirus to transduce cells; biotinylated particles can be bound to streptavidin coated supports preserving viral infectivity. Nanostructured surfaces have been shown to promote cell and protein adhesion (6, 7). In particular, titanium dioxide (TiO_2) is known as a biocompatible material and it is widely use in clinical devices (8). We have deposited by supersonic cluster beam deposition (9) a nanostructured TiO_2 film (ns-TiO_2) on slide and created, by robotic spotting, an array of streptavidin-biotinylated viral vectors. We obtained site-localized overexpression and downregulation, by RNA interference, of specific genes. Our ns-TiO_2 retroviral array is a novel powerful tool for the study of gene function of family of genes and the identification of drug targets in normal primary cells and their pathological counterparts.

References

(1) Ziauddin and Sabatini, Nature. 2001; 411(6833): 107–10.
(2) Wu RZ. et al., Trends Cell Biol. 2002; 12(10): 485–8.
(3) Mousses S. et al., Genome Res. 2003; 13(10): 2341–7.
(4) Kumar R. et al., Genome Res. 2003; 13(10): 2333–40.
(5) Hughes C. et al., Mol. Ther. 2001; 3(4): 623—30.
(6) Zinger O. et al., Biomaterials. 2004; 25(14): 2695–2711.
(7) Yahyapour N. et al., Biomaterials. 2004; 25(16): 3171–6.
(8) Ulrike Diebold, Surf. Sci. Rep. 2003; 48(5–8): 53–229.
(9) Barborini E. et al., Appl. Phys. Lett. 2002; 81(16): 3052–3054.

Tumor-Infiltrating B Lymphocytes as Efficient Source of Highly Selective Immunoglobulins Recognizing Tumor Cells

Emiliano Pavoni[1], Giorgia Monteriu[1], Daniela Santapaola[1], Fiorella Petronzelli[2], Anna Maria Anastasi[2], Rita De Santis[2], Olga Minenkova[1]

[1] Kenton Labs, c/o Sigma-Tau, Pomezia (RM), Italy
[2] Sigma-Tau, Pomezia (RM), Italy

Significance of spontaneous humoral response in cancer patients was investigated through analysis of oligoclonality of antibodies derived from tumor-infiltrating lymphocytes. Capacity of patient's immune system to produce tumor-specific antibodies within of tumor tissue was explored by construction of recombinant scFv antibody libraries derived from tumor-infiltrated B-lymphocytes from breast cancer patients. We panned the libraries against three available surface tumor antigens, as well as performed selection with intact breast carcinoma cells. A panel of novel recombinant cancer-specific antibodies was isolated from the libraries. In this study, we demonstrate that tumor samples obtained as discarded surgical material can be used as appropriate source for construction of recombinant phage display libraries enriched for tumor-specific antibodies. Isolation of a panel of anti-tumor scFvs toward known surface antigens, as well as to breast carcinoma living cells, shows this approach to be very promising in developing the tumor-targeting antibodies potentially useful for therapy and diagnosis of cancer. Anticancer antibodies, recognizing living cancer cells may lead to the discovery of new tumor surface antigens.

Role of Tip60 in Cell Cycle Checkpoints

Samantha Bennett, Anna De Antoni, Andrea Musacchio, and Bruno Amati

Department of Experimental Oncology, European Institute of Oncology, Milan, Italy

The histone acetyl-transferase Tip60 is a co-factor for a variety of transcriptional activators, including Myc, E2F and p53. Tip60 also has direct functions in the DNA damage response (DDR), in particular through the acetylation of ATM and H2AX. Work in our lab shows that Tip60 is critical for oncogene-induced DDR and tumor suppression in E?-myc mice (see abstract by Gorrini et al.). Here, we report molecular interactions that suggest an additional function for Tip60 in the spindle assembly checkpoint (SAC). We show that in co-transfected HeLa cells, Tip60 associates with Mad2 and p31COMET, two important regulators of the SAC. We are currently investigating whether these proteins are substrates for Tip60 and whether Tip60 regulates the SAC. Our progress will be presented at the meeting.

Biosynthesis of Estradiol in Normal Breast and Tumor Tissues

Robert T. Chatterton, Jr., and Villian Naeem

Department of Obstetrics and Gynecology, Northwestern University, Chicago, USA

17beta-hydroxysteroid dehydrogenases (17beta-HSD), and the impact of estrogen sulfates (ES) on availability of estradiol (E2) to normal and tumor tissues were studied in postmenopausal women. Enzyme activities were measured using tritiated-steroid precursors. Tissues were obtained during surgery for invasive ductal carcinoma.

Results: The median values for E1 formation from E1S were 17.2 and 6.6 pmol h^{-1} per million^{-1} cells in normal and tumor tissues, respectively. Sulfatase activities between normal and tumor tissues within individuals were correlated; $r = 0.52$ and 0.72 for E1 and E2 substrates, respectively. ES formation from E1 was 0.066 and 0.65 pmol h^{-1} per million cells in normal and tumor tissues, respectively. E2 formation from E1 was 0.23 and 2.10 pmol h^{-1} per million cells in normal and tumor tissues, respectively, and the 17beta-HSD activities between normal and tumor tissues within individuals were correlated ($r = 0.80$). E1 formation from E2 was high, 24.45 and 2.10 pmol h^{-1} per million cells in normal and tumor tissues, respectively.

Conclusions: (1) The 17beta-HSD reversible reactions between E1 and E2 were similar in tumor tissues (2.10 vs. 2.10) while in normal tissue, the oxidation reaction predominated (24.5 vs. 0.23). (2) 17beta-HSD but not sulfatase was rate limiting for formation of E2 in both normal and tumor tissues; sulfotransferase had little impact.

Che-1 Phosphorylation by ATM/ATR and CHK2 Kinases Activates P53 Transcription and the G2/M Checkpoint

Tiziana Bruno[1], Francesca De Nicola[1], Simona Iezzi[1], Daniele Lecis[2], Carmen D'angelo[3], Monica Di Padova[1], Nicoletta Corbi[4], Leopoldo Dimiziani[5], Laura Zannini[2], Christian Jekimovs[6], Marco Scarsella[3], Alessandro Porrello[7], Kum Kum Khanna[6], Marco Crescenzi[5], Carlo Leonetti[3], Silvia Soddu[7], Aristide Floridi[1], Claudio Passananti[4,8], Domenico Delia[2], and Maurizio Fanciulli[1,8].

[1] Laboratory, Department Of Therapeutic Programs Development, Regina Elena Cancer Institute, Via Delle Messi D'oro 156, 00158 Rome, Italy
[2] Department of Experimental Oncology, Istituto Nazionale Tumori, Via G. Venezian 1, 20133 Milan, Italy
[3] Experimental Chemotherapy Laboratory, Department of Experimental Oncology, Regina Elena Cancer Institute, Via Delle Messi D'oro 156, 00158 Rome, Italy
[4] Instituto Di Biologia E Patologia Molecolare, Cnr, Universita Di Roma La Sapienza, Piazzale Aldo Moro 5, 00185 Rome, Italy
[5] High Institute Of Health, Department of Environment and Primary Prevention, Viale Regina Elena 299, 00161 Rome, Italy
[6] Queensland Institute of Medical Research, P.O. Royal Brisbane Hospital, Brisbane, Queensland 4029, Australia
[7] Molecular Oncogenesis Laboratory, Department of Experimental Oncology, Regina Elena Cancer Institute, Via Delle Messi D'oro 156, 00158 Rome, Italy
[8] Rome Oncogenomic Center, Regina Elena Cancer Institute, Via Delle Messi D'oro 156, 00158 Rome, Italy

Che-1 is an RNA polymerase ii binding protein involved in the transcription of e2f target genes and induction of cell proliferation. Here we show that che-1 is involved in cellular response to DNA damage and that its downregulation induces a strong sensitivity to anticancer agents. We found that the checkpoint kinases atm/atr and chk2 physically and functionally interact with che-1 and promote its phosphorylation and accumulation. These che-1 modifications induce transcription of the *tp53* and *p21* genes by specific, phosphorylation-dependent recruitment of che-1 on their promoters. Finally, we found that the che-1-mediated transcriptional regulation is required for the dna-damage induced, p53- and p21-mediated maintenance of the g2/m checkpoint. These findings identify a new mechanism by which the checkpoint kinases atm/atr and chk2 regulate, via che-1, the p53-dependent responses to DNA damage.

Signaling to Actin Dynamics-Based Cell Migrations

Giorgio Scita

IEO, European Institute of Oncology, IFOM, The FIRC Institute for Molecular Oncology, Milan, Italy

Eps8-based signaling complexes: different facets of actin dynamics regulation. Dynamic assembly of actin filaments generates the forces supporting cell motility. Several recent biochemical and genetic studies have revealed a plethora of different actin binding proteins, whose coordinated activity regulates the turnover of actin filaments, thus controlling a variety of actin-based processes, including cell migration. Additionally, emerging evidence is highlighting a scenario whereby the same basic set of actin regulatory proteins is also the convergent node of different signaling pathways emanating from extracellular stimuli, like those from receptor tyrosine kinases. Within these pathways, RhoGTPases are crucial signal transducers. Here, emphasizing the role of a signaling protein, Eps8, we will discuss: how Eps8 and its family members can control the availability of free filaments barbed ends, by acting as barbed end capping proteins and the structure of the actin meshwork by cross-linking actin filaments. The integration and spatially restricted regulation of these activities is crucial for actin-based generation of forces leading to cell motility.

Raloxifene-Induced Myeloma Cell Apoptosis: A Study of NF-Kappa-B Inhibition and Gene Expression Signature

Sabine Olivier[1,2], Pierre Close[2], Emilie Castermans[3], Laurence de Leval[4], Sebastien Tabruyn[5], Alain Chariot[2], Michel Malaise[1], Marie-Paule Merville[2], Vincent Bours[2]*, and Nathalie Franchimont[1]*.

[1] Departments of Rheumatology
[2] Clinical Chemistry and Human Genetics
[3] Hematology
[4] Pathology and
[5] Genetic Engineering
* Center for Biomedical Integrative Genoproteomics, University of Liège, CHU Sart-Tilman 4000 Liège, Belgium:
The two senior authors share equal responsibility.

As multiple myeloma (MM) remains associated with a poor prognosis, novel drugs targeting specific signaling pathways are needed. In the present report, we studied the antitumor activity of raloxifene, a selective estrogen receptor modulator (SERM), on MM cell lines. Raloxifene effects were assessed by MTS reduction assay, cell cycle analysis and western blotting. Mobility shift assay, immunoprecipitation, ChIP assay and gene expression profiling were performed to characterize the mechanisms of raloxifene-induced activity. Indeed, raloxifene, as well as tamoxifen, decreased JJN-3 and U266 myeloma cell viability and induced caspase-dependent apoptosis. Moreover, raloxifene inhibited constitutive NF-kappaB activity in myeloma cells by removing p65 from its binding sites through ERalpha interaction with p65. Importantly, micro-array analysis showed that raloxifene treatment decreased the expression of known NF-kappaB-regulated genes involved in myeloma cell survival and myeloma-induced bone lesions (e.g. c-myc, mip-1alpha, hgf, pac1,...) and induced the expression of a subset of genes regulating cellular cycle (e.g., p21, gadd34, cyclin G2, . . .). In conclusion, raloxifene induces myeloma cell cycle arrest and apoptosis partly through NF-kappaB-dependent mechanisms. These findings also provide a transcriptional profile of raloxifene treatment on MM cells offering the framework for future studies of SERMs therapy in multiple myeloma.

Dissecting the Biological Functions of *Drosophila* Histone Deacetylates by RNA Interference and Transcriptional Profiling

Cristiana Foglietti, Gessica Filocamo, Enrico Cundari[§], Emanuele De Rinaldis, Armin Lahm, Riccardo Cortese, and Christian Steinkuhler*

Istituto di Ricerche di Biologia Molecolare "P. Angeletti"- IRBM- Merck Research Laboratories Rome, Italy and
[§] *Consiglio Nazionale delle Ricerche, Rome, Italy*

Histone deacetylase inhibitors have entered clinical trials for several malignancies. These compounds target up to 11 members of the HDAC family and it is not clear which HDACs are responsible for the antiproliferative activity. To get more insight into the functions of the individual HDAC family members we have used RNAi in combination with microarray analysis in *Drosophila* S2 cells. Silencing of *Drosophila* HDAC1 (DHDAC1) lead to increased histone acetylation, cell growth inhibition and deregulated transcription. RNAi of DHDAC3 also perturbed transcription but resulted in only a modest growth phenotype. Silencing of DHDAC2 lead to increased tubulin acetylation levels but was not associated with a deregulation of gene expression. No growth phenotype and no significant deregulation of gene expression was observed upon silencing of DHDACs 4 and X. Loss of DHDAC1 or exposure of S2 cells to the small molecule HDAC inhibitor Trichostatin both lead to a G2 arrest and were associated with overlapping gene expression signatures. A large number of these genes were shown to be also deregulated upon loss of the co-repressor SIN3. We conclude that: (1) DHDACs 1 and 3 have both overlapping and distinct functions in the control of gene expression. (2) Under the tested conditions, DHDACs 2, 4, and X have no detectable transcriptional functions in S2 cells. (3) The antiproliferative and transcriptional effects of Trichostatin are largely recapitulated by the loss of DHDAC1. (4) The deacetylase activity of DHDAC1 significantly contributes to the repressor function of SIN3.

A Persistent DNA-Damage Response in Senescent Human Fibroblasts

Marzia Fumagalli and Fabrizio d'Adda di Fagagna

IFOM, Via Adamello 16, 20139 Milan, Italy

Cellular senescence is a barrier to unruly cellular proliferation generating a robust DNA damage response, triggered by telomere dysfunctions. We are probing into the mechanisms that trigger a DNA damage response in human primary fibroblasts bearing critically short telomeres. We knocked down transiently (siRNA interference) or permanently (by the expression of short hairpin, RNAs delivered by lentiviral vectors) several factors involved in the DNA-damage response into senescent human fibroblasts: this approach allows us to study whether a checkpoint inactivation allows cells to re-enter into S-phase and to engage a productive cytokinesis. We are also presently monitoring the transformation potential of permanently knocked-down cells. Moreover, we are studying whether senescence is enforced by the continuous activity of checkpoint factors or it is established by an initial burst of their activity without their further involvement. We discovered that DNA-damage response factors are still engaged years after the initial establishment of cellular senescence. Nevertheless, senescent cells are DNA-repair proficient. We conclude that SDFs (Senescence associated DNA damage foci) are irreparable double-strand breaks, which persist for very long time. We are presently studying the peculiar characteristics of these irreparable breaks.

Role of MicroRNAs in Myeloid Differentiation

Rosa Alessandro, Ballarino Monica, De Angelis Fernanda Gabriella, Incitti Tania, Sthandier Olga, Fatica Alessandro, Bozzoni Irene.

Department of Genetics and Molecular Biology and IBPM, Institute Pasteur Cenci-Bolognetti, University of Rome "La Sapienza", P.le A. Moro 5, 00185 Rome, Italy

MicroRNAs (miRNAs) represent a new class of small noncoding RNAs, known to be negative regulators of gene expression at a post-transcriptional level. In the last few years, a key role in development and differentiation has been demonstrated for miRNAs in plants and animals and some of them have also been correlated with cancer. Our research project focuses on the identification of miRNAs specifically expressed in different hematopoietic lineages and on the study of their function. We utilize promielocytic leukemia cell lines, which can be induced to differentiate into granulocytes or monocytes/macrophages by retinoic acid (RA) or TPA treatment, respectively. In a previous work (Fazi et al., 2006), we demonstrated that a minicircuitry involving a miRNA (miR-223) and two well-known transcriptional factors (C/EBPalpha and NFI-A) determinates the cell fate during granulopoiesis. At present, we intend to characterize the whole set of microRNAs involved in myeloid differentiation, their target genes and the molecular circuitries in which they are involved. This study will provide a new insight into the understanding of the molecular processes required for hematopoietic differentiation and the possibility of identifying new therapeutic targets in leukaemia.

A Role of Angiomotin in Endothelial Cell–Cell Contacts

Indranil Sinha, Anders Bratt, and Lars Holmgren

Department of Oncology-Pathology, Cancer Centrum Karolinska, Karolinska Institute, R8:03 Karolinska University Hospital, 171 76 Stockholm, Sweden

An attractive approach to control angiogenesis is to use the molecules that are normally present in our bodies and that negatively control vessel growth. One of the first inhibitor of this kind to be reported was angiostatin. The results of angiostatin in tumor models were very promising; angiostatin could maintain tumors in a state of no growth! However, the half-life of angiostatin when injected into cancer patients was less than 3 h making it extremely difficult to deliver enough to maintain an effective dose over time. To circumvent this problem our group has identified a protein that mediates the effect of angiostatin. We named this molecule angiomotin (angio = vessel, motin = motility) as it affects the movement of the cells that are part of the vessel wall. Here we present data showing that angiomotin also plays a role in cell junction formation.

Immunofluorescence analysis shows that both the angiomotin isoforms, p80 and p130, localize to cell–cell contacts in vitro and in vivo. Western blot and immunofluorescence studies showed that angiomotin protein levels are upregulated when cell contacts are formed. Co-immunoprecipitation study showed the longer angiomotin isoform, p130, to be co-precipitated with the tight junction protein MAGI-1b. Paracellular permeability, as measured by diffusion of fluorescein isothiocyanate-dextran, was reduced by p80 and p130 angiomotin expression with 70% and 88%, respectively, compared with control. Angiostatin did not have any effect on cell permeability but inhibited the migration of angiomotin-expressing cells in the Boyden chamber assay. We conclude that angiomotin, in addition to controlling cell motility, may play a role in the assembly of endothelial cell–cell junctions.

Epigenetic Determinants of MYC Binding to the Human Genome

Ernesto Guccione[1], Francesca Martinato[1], Lucilla Luzi[1,2], Giacomo Finocchiaro[1,2], Laura Tizzoni[3], Valentina Dall' Olio[3], Loris Bernard[1,3], and Bruno Amati[1,4]

[1] *Department of Experimental Oncology, European Institute of Oncology (IEO)*
[2] *FIRC Institute of Molecular Oncology (IFOM)*
[3] *Real-Time PCR Facility*
[4] *IFOM-IEO Campus, Milan 20139, Italy*

Large-scale chromatin immunoprecipitation (ChIP) studies are unraveling the distribution of transcription factors (TFs) along eukaryotic genomes, but specificity determinants remain elusive. Gene regulatory regions display distinct histone variants and covalent modifications (or marks). The histone code" hypothesis posits that these marks – or combinations thereof – modulate protein recognition, but whether this applies to TFs remains unknown. Based upon large-scale datasets and quantitative ChIP, we dissect the correlations between 35 histone marks and genomic binding by the TF Myc. Our data reveal a relatively simple combinatorial organization of histone marks in human cells, with few main groups of marks clustering on distinct promoter populations. A stretch of chromatin bearing high H3 K4/K79 methylation and H3 acetylation (or "euchromatic island") is a strict pre-requisite for recognition of any target site by Myc (whether the canonical consensus site CACGTG or alternative sequences). These data imply that selective tethering of a TF to restricted chromatin domains is rate limiting for sequence-specific DNA binding in vivo.

Different Activation Profiles of MAPK/EPK1/2 Mediate Opposite Effects of the Neuropeptide Y–Y1 Receptor System on Prostate Cancer Cell Growth

Massimiliano Ruscica, Elena Dozio, Marcella Motta, Paolo Magni

Center for Endocrinological Oncology, Department of Endocrinology, University of Milan, Milan, Italy

The pleiotropic neuropeptide Y (NPY) is a neuroendocrine molecule that could be play a significant role in the progression of human prostate cancer (PCa). This study evaluated the direct effect of NPY on the growth of the human PCa cell lines LNCaP (androgen dependent), DU145 and PC3 (androgen independent). The Y1 receptor (Y1-R) was expressed in all PCa cell lines. Treatment with 10–8 M NPY reduced the proliferation of LNCaP and DU145 cells and increased that of PC3 cells. The specific Y1-R antagonist BIBP3226 (10–6 M) abolished such effects, suggesting a mandatory role of Y1-R in this process. LNCaP cells showed elevated constitutive levels of phosphorylated extracellular kinase (ERK)1/2, which were not affected by NPY. In DU145 cells, treatment with 10–8 M NPY stimulated a long-lasting (>6 h) ERK1/2 phosphorylation, whereas, in PC3 cells, this effect was rapid, transient and mediated by protein kinase C. In DU145 and PC3 cells, NPY-induced ERK1/2 phosphorylation was prevented by a pretreatment with BIBP3226, further suggesting the involvement of Y1-R. Treatment with 10-8 M NPY reduced forskolin-stimulated cAMP accumulation only in PC3 cells. In conclusion, Y1-R activation by NPY represents an important regulator of the proliferation of different PCa cell lines.

Expression and Function of CX3CR1 Receptor for Fractalkine on Pancreatic Tumor Cells its Involvement in Tumor Dissemination

Marchesi F[1], Monti P[2], Leone BE[3], Piemonti L[2], Mantovani A[1,4], Allavena P[1]

[1] Department Immunology Istituto Clinico Humanitas, Rozzano, Italy
[2] JDRF Telethon for Beta Cell Replacement, S. Raffaele Scientific Institute, Milan, Italy
[3] University of Milan Bicocca, Italy
[4] Institute of Pathology, University of Milan, Italy

In this study, we have analyzed the presence of the chemokine receptor CX3CR1 on 11 pancreatic tumor cell lines and tumor cells from surgical samples of patients with pancreatic adenocarcinoma. Six of eleven pancreatic tumor cell lines express significantly higher amounts of CX3CR1 transcripts, compared to a cell line derived from normal ductal epithelium used as reference. CX3CR1 is also expressed in 7/7 primary tumor cells isolated from surgical specimens (>95% cytokeratine-7 positive). Flow cytometric analysis with a specific mAb confirmed these results at the protein level. We also tested the functional activity of CX3CR1-positive tumor cells. AsPC1 and A8184 dose-dependently migrated in response to the specific ligand CX3CL1 (Fractalkine) in classical chemotaxis assays and this effect was blocked by specific anti-CX3CR1 antibodies. CX3CL1 also functions as an adhesion molecule. CX3CR1–positive pancreatic tumor cells showed enhanced adhesion to CX3CL1-coated plastic as well as to neuroblastoma cells (SKN-BE) which produce and release CX3CL1 upon stimulation with TNFα/IFNγ. Pancreatic tumor cell adhesion was specifically inhibited by antibodies anti-CX3CL1. Overall these results demonstrate that functional Fractalkine receptor is expressed in pancreatic cancer cells and could be involved in tumor dissemination.

Genome-Wide Chromatin-Based Isolation of Active cis-Regulatory DNA Elements from Human Cells

Gargiulo G[2], Urnov F[1], Levy S[3], Ballarini M[2], Bucci G[2], Romanenghi M[2], Pelicci PG[2], Gregory PD[1], and Minucci S[2]

[1] *Sangamo BioSciences, Inc., Pt. Richmond, CA 94804*
[2] *European Institute of Oncology, Milan, Italy; Department of Biomolecular Sciences (SM, MB, GG, GB) and School of Medicine (PGP), University of Milan.*
[3] *J. Craig Venter Institute, 9704 Medical Center Drive, Rockville, MD 20850, USA*

Development, cell proliferation, and cellular responses to environmental signals are all orchestrated by coordinated patterns of gene expression. DNA-encoded genetic information is packaged into a chromatin polymer, which must be opened to allow increased accessibility for gene regulatory factors, or compacted to restrict access of the transcriptional machinery to target genes. The study of changes in chromatin remodeling at *cis*-regulatory elements can be a powerful tool to understand normal and diseased cell physiology. A number of genome-wide technologies have been developed to identify the location of gene regulatory elements, such as sequence conservation, ChIP-chip and computational analysis. Since each method has its own set of disadvantages, a combination of different techniques will likely be needed to successfully discovery all gene regulatory elements. We describe a rapid, simple, generally applicable method to identify regulatory DNA elements in human cells genome-wide: mild treatment of cells with a restriction enzyme followed by the isolation of the DNA ends generated by the nuclease in chromatin. Either microarray hybridization or high-throughput sequencing can be used with confidence as 90% of the fragments we isolated this way correspond to bona fide regulatory DNA. In all cell types we studied the majority of such elements did not coincide with core promoters of annotated genes.

The Contribution of Variant Hepatocyte Nuclear Factor 1 and E-Cadher into ovarian carcinoma Cell Growth

Giuseppina De Santis, Giancarlo Castellano, Mimma Mazzi, Silvana Canevari, Antonella Tomassetti

Unit of Molecular Therapies, Department of Experimental Oncology, Istituto Nazionale Tumori, Milan, Italy

The cell monolayer of the ovarian surface epithelium (OSE) gives rise to over 80% of human epithelial ovarian carcinomas (EOCs). At the first stages of tumorigenesis, OSE show an aberrant commitment to epithelial differentiation with expression of the specific epithelial marker E-cadherin. We have also found that the EOCs express another epithelial marker, the Variant Hepatocyte Nuclear Factor 1 (vHNF1), a homeodomain-containing transcription factor initially identified by activation of genes responsible of liver differentiation. We are examining in EOC cells (1) the relationship between the expression of the 2 epithelial markers E-cadherin and vHNF1, and (2) whether E-cad might contribute in the regulation of signal other than adhesion in ovarian cancer cells. Knockdown by small interfering RNA to vHNF1 in the ovarian carcinoma IGROV1 and SKOV3 cells lead to downregulation of E-cadherin expression and of AKT phosphorylation, with no effect on MEK/ERK and beta-catenin/TCF pathways. Accordingly, PI3K/AKT pathway was affected upon knockdown of E-cadherin. Furthermore, the E-box binding protein SNAIL appeared to be not responsible for downregulation of E-cadherin expression. We are now transfecting the relevant cDNAs in immortalized OSE cells that will be also analyze for phenotypic and signaling pathway changes.

Replication Fork Progression is Impaired by Transcription in Hyper-Recombinant Yeast Cells Lacking a Functional THO Complex

R.E. Wellinger, F. Prado, and A. Aguilera

Universidad de Sevilla, Dpt. de Genetica, Sevilla, Spain

THO/TREX is a conserved eukaryotic protein complex operating at the interface between transcription and messenger ribonucleoprotein (mRNP) metabolism. THO/TREX complex has been shown to be important for proliferation of human cells. Additionally, high expression of the human homologue of Hpr1 (hTREX84) has been linked to metastatic breast cancer. THO mutations impair transcription and lead to increased transcription-associated recombination (TAR). These phenotypes are dependent on the nascent mRNA; however, the molecular mechanism by which impaired mRNP biogenesis triggers recombination in THO/TREX mutants is unknown. We provide evidence that deficient mRNP biogenesis causes slow-down or pausing of the replication fork in hpr1δ mutants. Impaired replication appears to depend on sequence-specific features since it was observed upon activation of lacZ but not leu2 transcription. Replication fork progression could be partially restored by hammerhead ribozyme guided self-cleavage of the nascent mRNA. Additionally, hpr1δ increased the number of cells containing recombinational (Rad52) repair foci, and the number of S-phase-dependent, but not G2-dependent, TAR events. Our results link transcription-dependent genomic instability in THO mutants with impaired replication fork progression, suggesting a molecular basis for a connection between inefficient mRNP biogenesis and genetic instability.

Investigation of the Molecular Mechanisms Regulating the G2 Checkpoint Focusing on Chk1

Laura Carrassa, Giovanna Damia, Monica Ganzinelli, and Massimo Broggini

Molecular Pharmacology Laboratory, Department of Oncology, Istituto di Ricerche Farmacologiche Mario Negri, Milan, Italy

Following DNA damage, checkpoints pathways are activated in the cells to halt the cell cycle thus ensuring repair or inducing cell death. Chk1 and Chk2 protein kinases play a crucial role in the G2 checkpoint. We previously downregulated by siRNA the expression of Chk1 and/or Chk2 in the HCT-116 colon carcinoma cell line and in its subclones with either p53 or p21 inactivated by homologous recombination, finding out that inhibiting Chk1 but not Chk2 caused a greater abrogation of the G2 block and a greater sensitization to different DNA damaging agents in the cell lines with a defective G1 checkpoint than in the parental wild-type cells. These data emphasize the role of Chk1 as a molecular target to inhibit in tumors with a defective G1 checkpoint to increase the selectivity of anticancer treatments. Tetracycline-inducible expressing Chk1 siRNA clones have been obtained transfecting both HCT-116 wild-type and p53 deficient cells with a plasmid inducible expressing Chk1 siRNA, showing a strong inhibition of Chk1 protein levels following tetracycline induction. These clones have been transplanted in nude mice and experiments are undergoing to verify the downregulation of Chk1 upon addition of tetracycline in their drinking water.

Interplay Between Notch and the Histone Methyl Transferase TRX in Tumorigenesis

Esther Caparros, Irene Gutierrez-Garcia, and Maria Dominguez

Instituto de Neurociencias CSIC-UMH, Departamento de Neurobiología del Desarrollo. Campus de San Juan, Alicante, Spain

A common trait of human cancers is the inactivation of tumor-suppressor genes by epigenetic processes, particularly via the modifications of histones and DNA hypermethylation. How are these damaging epigenetic changes initiated in cells that become precursors of cancer? And which is the connection between the epigenetic processes and the developmental pathways controlling cell proliferation? Recently, by studying tumorigenesis in the *Drosophila* eye (Ferres-Marco et al., Nature 2006), we have identified two epigenetic silencers, Pipsqueak and Lola, that participate in this process. The dysregulation of Pipsqueak and Lola coupled with Notch hyperactivation induces the formation of metastatic tumors. These tumors depend on the histone deacetylase Rpd3 and the histone methyltransferases (HMTs), Su(var)3-9 and E(z). Associated with the development of these tumors, we observed the depletion of the histone H3 lysine (K) 4 methylation, an epigenetic tag of active chromatin, which depends on trithorax (trx) HMT. Intriguingly, our analyses of the role of trx in tumorigenesis strongly suggest that this histone methyltransferase Trx acts as a tumor-suppressor. Given the high prevalence of acute leukaemia associated with chromosomal translocations at the MLL/TRX loci, we believe that these discoveries may pave the way for new research in Notch- and Trx-associated cancers.

Somatic Mutations of Epidermal Growth Factor Receptor Signal Transducers in Cholangiocarcinoma

Ymera Pignochino[1]*, Junia Yara Penachioni[3]*, Carlo Zanon[3], Giuliana Cavalloni[1], Francesco Leone[1], Ivana Sarotto[2], Mauro Risio[2], Alberto Bardelli[3], Massimo Aglietta[1]

[1] *Division of Clinical Oncology, University of Torino Medical School, Institute for Cancer Research and Treatment (IRCC), Candiolo, Italy*
[2] *Unit of Pathology University of Torino Medical School, Institute for Cancer Research and Treatment (IRCC), Candiolo, Italy*
[3] *Oncogenomics Center, University of Torino Medical School, Institute for Cancer Research and Treatment (IRCC), Candiolo, Italy*
* These authors equally contributed to this work

Somatic mutations in the tyrosine kinase (TK) domain of *EGFR* gene in lung cancer predict TK inhibitors sensitivity (TKIs). Nonetheless, it is not the sole factor, genes downstream of EGFR signaling, may be involved in tumorigenesis and in response to TKIs. We previously described somatic mutations of *EGFR* gene in cholangiocarcinoma. For optimal therapeutic approach a comprehensive understanding of EGFR signaling is required. We analyzed 38 paraffin-embedded samples of cholangiocarcinoma by DNA sequencing of exon 9 and 20 of PI3K, exon 2 of KRAS and exon 15 of BRAF which are frequently mutated in many tumors. Of the 38 specimens, 5 (13.2%) had hotspot mutations in PI3K (codon 545, 546, 1,048 and 1,059); 3 (7.9%) had G13D mutation in KRAS; 4 (10.5%) had V599E mutation in BRAF. In some sample, mutations of multiple trasducers were present simultaneously: one patient had mutated PI3K and KRAS, a second had mutated EGFR and PI3K, a third had mutated EGFR, PI3K and BRAF. This finding suggests that an accurate analysis may be done before treatment with TKIs, as the presence of activating mutations in EGFR signal transducers can abrogate the TKIs responsiveness due to EGFR–TK mutations.

Role of Myc in Transcriptional Responses Induced by a Variety of Key Signaling Pathways

Daniele F. Perna[1], Alison P. Smith[1], Remo Sanges[2], Elia Stupka[2], Andreas Trumpp[3], and Bruno Amati[1]

[1] European Institute of Oncology, Via Ripamonti 435, Milan Italy
[2] Telethon Institute of Genetics and Medicine (TIGEM), Via Pietro Castellino 111, Naples Italy
[3] Swiss Institute for Experimental Cancer Research (ISREC), Epalinges, Switzerland

The Myc oncoprotein is a transcription factor that can either activate or repress gene expression. High-throughput analyses of gene expression have failed to reveal any universal signature for Myc-regulated transcription. Furthermore, it has been suggested that Myc is required but not sufficient for gene activation and that other signals must be involved. Thus, Myc may function as a "transcriptional switch" which modifies the state of responsiveness of many genes to other signals. The aim of this project is to challenge this hypothesis by testing if Myc cooperates with a variety of key signaling pathways (TGF-beta, TNF-alfa and serum response). Using Affymetrix microarrays, we are currently collecting profiling data in 3T9f/f fibroblasts, derived from conditional "flox-myc" mice, in which c-myc has been either deleted or not with a stably expressed, OHT-inducible form of Cre. Preliminary data show that the majority of genes responsive to those pathways remain functional upon Myc deletion. Using clustering analyses we are looking for genes that behave differentially in the presence and absence of Myc. Our progress in analyzing these transcriptional responses will be presented.

Molecular Signature of Malignant Cutaneous Melanoma and BRAF Mutation

Maria Scatolini, Maurizia Mello Grand, Francesco Acquadro, Francesca Guana and Giovanna Chiorino (Fondo Edo Tempia Biella), Tiziana Venesio (IRCC- Candiolo, Torino)

In our study we analyzed 55 biopsies from common nevi ($n = 22$), primary radial growth phase malignant melanoma ($n = 15$), primary vertical growth phase malignant melanoma ($n = 13$) and melanoma metastasis ($n = 5$). Global gene expression profiling of the tissues was performed using whole genome oligo-swap microarrays with a dye-swap duplication scheme. All the samples were also sequenced to look for BRAF mutations. The first goal of our study was to identify candidate genes of melanoma progression and candidate markers of melanoma metastasis. The second goal was to look for any correlation between BRAF mutation and patient phototype or sun exposure in nevi and melanomas. The third one was to compare the gene expression profiling of nevi and melanomas with or without V599E mutation on BRAF exon 15 to find any alternative way of melanoma development not involving the MAPK cascade signaling. Results will be presented and discussed.

Novel Chimpanzee Serotype-Based Adenoviral Vector as Vaccine for CEA

Daniela Peruzzi, Barbara Cipriani, Agostino Cirillo, Bruno Ercole Bruni, Annalisa Meola, Alfredo Nicosia, Riccardo Cortese, Gennaro Ciliberto, Stefano Colloca, Nicola La Monica, and Luigi Aurisicchio

Istituto di ricerche di Biologia Molecolare P. Angeletti (IRBM) Via Pontina Km 30.600. 00040- Pomezia, Roma, Italy

Adenovirus (Ad) is an attractive vector for genetic vaccination due to its ability to infect several tissues. This property is impaired by pre-existing immunity to the human adenoviruses in most humans. Therefore finding uncommon alternative Ad serotypes becomes a priority. We have used a chimpanzee Adeno serotype 3 (E1,E3 deleted, ChAd3) engineered to express the human carcinoembryonic antigen (CEA) protein to vaccinate CEA. Tg mice in which human CEA shows a tissue distribution similar to that of humans. Animals were intramuscularly immunized with ChAd3-CEA and expression of human CEA was monitored in the serum. Development of a CEA-specific CD8+ T cell response was observed after a single injection and was comparable, in terms of efficacy and kinetics, to that of mice immunized with a similar vector based on human serotype 5 (Ad5-CEA) expressing CEA. Importantly, the capability of ChAd3-CEA to elicit an immune response against CEA was not abrogated in animals pre-exposed to human weight Ad5 but worked as booster of the immunity elicited even by other Ad vectors. In conclusion, our data show that this novel chimp Ad based vaccine vector is immunogenic, breaks tolerance to a tumor antigen and may overcome pre-existing immunity phenomena in humans.

JAM-A Deficient Polymorphonuclear Cells Show Reduced Diapedesis in Peritonitis and Heartischemia-Reperfusion Injury Mice

M. Corada[*,†], S. Chimenti[†], M.R. Cera[*], M. Vinci[*], M. Salio[†], F. Fiordaliso[†], N. De Angelis[†], A. Villa[‡], M. Bossi[‡], L.I. Staszewsky[†], A. Vecchi[†], D. Parazzoli[§], T. Motoike[∂], R. Latini[†∂∂], and Elisabetta Dejana[*,†,#]

[*] FIRC Institute of Molecular Oncology, 20139 Milan, Italy
[†] Mario Negri Institute for Pharmacological Research, 20157 Milan, Italy
[‡] Consorzio MIA and Department of Neuroscience, University of Milano-Bicocca, 20052 Monza, Italy
[§] IEO European Institute of Oncology, 20141 Milan, Italy
[∂] The University of Texas Southwestern Medical Center at Dallas, Dallas, Texas, USA
[∂∂] Department of Medicine, New York Medical College, Valhalla NY, USA
[#] Department of Biomolecular and Biotechnological Sciences, School of Sciences, University of Milan, 20100 Milan, Italy

Junctional adhesion molecule-A (JAM-A) is a transmembrane adhesive protein expressed at endothelial junctions and in leukocytes. Here we report that JAM-A is required for the correct infiltration of polymorphonuclear leukocytes (PMN) into an inflamed peritoneum or in the heart upon ischemia-reperfusion injury. The defect was not observed in mice with an endothelium-restricted deficiency of the protein, but was still detectable in mice transplanted with bone marrow from JAM-A$^{-/-}$ donors. Microscopic examination of mesenteric and heart microvasculature of JAM-A$^{-/-}$ mice showed high numbers of PMN adherent on the endothelium or entrapped between endothelial cells and the basement membrane. In vitro, in the absence of JAM-A, PMN adhered more efficiently to endothelial cells and basement membrane proteins and their polarized movement was strongly reduced. This work describes a novel and nonredundant role of JAM-A in controlling PMN diapedesis through the vessel wall.

Molecular Profiling Identifies Deregulated Expression of Multiple ETS Family Transcription Factors in Prostate Cancer

Maurizia Mello Grand, Maria Scatolini, Francesco Acquadro and Giovanna Chiorino (Fondo Edo Tempia- Biella, ITALY), Veronica Albertini, Afua Mensah, Carlo V Catapano, Giuseppina M Carbone (IOSI - Bellinzona, Switzerland)

Prostate cancer is one of the most common neoplasia in men. This cancer is clinically and genetically heterogeneous and varies in his biological aggressiveness. We performed gene expression profiling on 93 human prostate samples including 17 normal biopsies, 67 clinically localized adenocarcinomas of the peripheral zone of the gland and 9 benign prostatic hyperplasia (BPH). Comparative hybridization of tissues against a commercially available normal prostate reference was realized using oligonucleotide glass arrays with sequences representing over 18,000 well-characterized human genes, in a dye-swap duplication scheme. The aims of the study were: to identify genes involved in carcinogenesis; to improve the histological classification with molecular data; to discover diagnostic and prognostic biomarkers to detect and/or anticipate the clinical evolution of the cancer by correlating gene expression profiles with follow-up data. We also focused on a family of transcription factors whose activation or inhibition could have an important role in the development of the disease. Gene profiling of tumors was matched with histopathological and clinical data and aberrant expression of several ETS family genes was identified at distinct disease stages and patient subgroups. Collectively, molecular profiling suggested a potential role of multiple ETS members in prostate cancer pathogenesis.

HDAC1/2 Interacting Protein MTA2 is an Activating Cofactor of their Enzymatic Activity

A. Lahm, C. Paolini, C. Nardi, M. Pallaoro, C. Steinkuhler, and P. Gallinari

Istituto di Ricerche di Biologia Molecolare (I.R.B.M) P. Angeletti, Pomezia (Rome)

The acetylation state of histones and several nonhistone proteins is regulated by the opposite action of histone acetyltransferases (HATs) and histone deacetylases (HDACs) that play a critical role in transcription, cell proliferation, differentiation, genome stability, and stress response. An imbalance of these reactions leads to aberrant cell behavior and neoplastic transformation. HDACs are overexpressed or aberrantly recruited in different human malignancies and several HDAC inhibitors have entered clinical trials as antitumor agents. HDACs work within large multiprotein complexes. HDAC3 enzymatic activity strictly requires the interaction with SMRT/N-CoR corepressors' DAD, including a DAD-specific motif followed by a SANT domain. At least one SANT domain is also present in three HDAC1/2 corepressors, MTA, CoREST, and MI-ER1. Here we show that by co-purifying recombinant HDAC1 or HDAC2 with MTA2 their deacetylase activity in vitro is significantly enhanced. In all abovementioned HDAC1/2 corepressors sequence homology extends N-terminal to the SANT domain, allowing the definition of a DAD-homology region as in SMRT/N-CoR. MTA and MI-ER1 share an additional, more N-terminal domain, ELM2. By deletion mutagenesis of MTA2, a region encompassing both ELM2 and DAD is shown to be necessary and sufficient for the interaction with HDAC1 and is the minimal region capable of increasing HDAC1 enzymatic activity in vitro and transcriptional repression potential in cells.

The Monouniquitin Network at Work

Simona Polo

IFOM, the FIRC Institute for Molecular Oncology Foundation, Milan, Italy

The Ub modification regulates protein stability and function, with impact on, virtually, every major cellular phenotype. One paramount function of Ub is to determine proteolysis of intracellular proteins. In this process, an Ub chain, composed of at least four Ubs branching from Lys48, is appended to target proteins that are then delivered to the proteasome for degradation. Nonproteolytic functions of Ub, conversely, rely on monomeric Ub or poly-Ub chains branched from lysines other than Lys48. In this case, ubiquitination functions as a regulatory modification that affects the structure, the activity, or the localization of the target protein. A series of ubiquitin recognition domains and motifs exists and are starting to be characterized. Thus, there are important similarities between the network of Ub-mediated interactions and the extensively characterized network of tyrosine phosphorylation. Our lab is interested in investigating how ubiquitination couples RTKs (receptor tyrosine kinases) with downstream pathways. Data regarding different on-going projects will be presented.

Telomerase Involvement in Gliobastoma Multiforme Angiogenesis

Maria Patrizia Mongiardi[1], Paolo Fiorenzo[1], Maria Laura Falchetti[1], Luigi Maria Larocca[2], Nicola Montano[3], Giorgio D'Alessandris[3], Giulio Maira[3], Roberto Pallini[3], Andrea Levi[1]

[1] *INeMM CNR Catholic University, Rome, Italy*
[2] *Institutes of Human Pathology Catholic University, Rome, Italy*
[3] *Neurosurgery, Catholic University, Rome, Italy*

Glioblastoma multiforme (GBM) is a very aggressive tumor with a prominent angiogenic phenotype. We previously demonstrated that endothelial cells of GBM vasculature express the catalytic subunit of telomerase, hTERT (1), and that GBM cells secrete factors able to induce hTERT expression and telomerase activity in primary human endothelial cells (HUVECs) (2). To understand the role of telomerase upregulation by the endothelial cells of the tumor vasculature, we engineered HUVECs to express either hTERT (hTERT-HUVECs), or a dominant negative hTERT (DN-hTERT-HUVECs), or short hairpin RNAs directed against hTERT mRNA (RNAi-hTERT-HUVECs). In vitro, engineered HUVECs did not differ from wild-type HUVECs in terms of response to cell stressing conditions, like serum deprivation or DNA damaging agents. However, when HUVEC cells were subcutaneously cografted with human GBM cells in nude mice, both DN-hTERT-HUVECs and RNAi-hTERT-HUVECs did not survive transplantation. Conversely, both hTERT-HUVECs and wild-type HUVECs, which expressed endogenous hTERT due to the GBM-mediated telomerase activation, survived for long periods and formed net-like structures that established anastomoses with the host vessels. This suggests that telomerase activity in endothelial cells may promote tumor angiogenesis.

(1) Pallini et al., J Neurosurg. 2001 94:961–71.
(2) Falchetti et al., Cancer Res. 2003 63:3750–4.

A New Role for the Bromodomain-Containing Protein brd7 in Regulating p53 Family Activities

Francesca Tocco[1], Fiamma Mantovani[1,2], Alessandra Rustighi[1], and Gianni Del Sal[1,2].

[1] *National Laboratory CIB, Padriciano, Trieste, Italy, and*
[2] *Department of Biochemistry, Biophysics and Chemistry of Macromolecules of the University of Triest, Triest, Italy*

The tumor suppressor p53 and its homologues p73 and p63 constitute a family of transcription factors that play key roles in maintaining genomic stability and cellular homeostasis. Despite the considerable knowledge on p53 activities, less is understood on the mechanisms that dictate the specificity of its response, while much less is known about the regulation of tumor suppressive activities of the other family members. The bromodomain containing protein brd7 has recently emerged as a common interactor of the p53 family members in a yeast two-hybrid screening performed in our lab. brd7 has been previously shown to bind acetylated histones through its bromodomain, and previous reports support a role for this factor in regulating transcription. These evidences make it an appealing candidate for modulating the transcriptional activity of p53 family proteins. In fact, we have observed that the ablation of brd7 leads to downregulation of p21 expression both in a p53 dependent and independent fashion. It is our interest to analyze brd7 functions further, in order to gain more insights on the mechanisms that govern the stress response of p53 family members, through regulating their recruitment to different promoters in response to specific stimuli.

Adherence Junctions Regulate TGF-b Signaling in Endothelial Cells

Noemi Rudini, Angelina Felici, and Elisabetta Dejana

FIRC-Institute of Molecular Oncology, Milano, Italy

Induction of angiogenesis is essential for tumor formation, growth, and dissemination. Transforming growth factor-β (TGF-b) secreted by cancer cells profoundly affects the tumor microenvironment with its immunosuppressive and angiogenic activities. In vascular endothelial cells (ECs), TGF-b mediates its cellular effects through the serine/threonine kinase receptors, TbRII and TbRI, and the coreceptor endoglin, coupled with Smad transcriptional regulators. Vascular endothelial (VE)-cadherin is expressed at interendothelial junctions of normal and tumor vessels referred to as adherens junctions (AJ). VE-cadherin transfers intracellular signals both through the association with VEGFR receptor thus enhancing its signaling and through the binding of catenins (a-, b-, p120) which, when released into the cytoplasm, translocate to the nucleus and regulate target gene transcription. The notion that TGF-b-mediated endothelial-to-mesenchymal transformation is impaired in b-catenin−/− ECs and that VE-cadherin KO strikingly resembles endoglin KO, prompted us to investigate a possible interplay between AJ and TGF-b pathway in ECs. To assess whether VE-cadherin/b-catenin signaling regulates TGF-b pathway, we performed Smad-dependent reporter gene assays in b-catenin+/+ vs b-catenin−/− ECs and in CHO vs VE+-CHO cells. The results show enhanced TGF-b/Smad reporter activity in sparse b-catenin−/− ECs and in confluent VE+-CHO cells, suggesting that clustered VE-cadherin/b-catenin signaling potentiates TGF-b-dependent transcriptional activity. Since AJ signaling impacts TGF-b pathway, we tested whether VE-cadherin physically interacts with TGF-b receptors. Briefly, VE-cadherin and TGF-b receptors have been transiently expressed into recipient cells and the co-immunoprecipitated complexes analyzed. The results show that the TGF-b receptors endoglin, TbRII, Alk1, and Alk5 associate with VE-cadherin in vivo. These findings demonstrate a functional crosstalk between TGF-b and AJ pathways providing prospects for better understanding the role of AJ proteins in TGF-b-mediated effects in angiogenic ECs.

Functions of AP-1 (c-Jun/c-Fos) in Liver Cancer Development

Lijian Hui, Harald Scheuch, and Erwin F. Wagner

Research Institute of Molecular Pathology (IMP), A-1030, Vienna, Austria

The activator protein 1 (AP-1) transcription factors can exert their oncogenic or tumor suppressive effects by regulating cell proliferation, apoptosis, differentiation and invasion. c-Jun antagonizes the pro-apoptotic activity of p53 in a diethylnitrosamine-phenobarbital (DEN-Pb) induced liver tumor model 1. Liver-specific ablation of c-Jun (c-jun$^{\Delta li*}$) at different stages of tumor development showed that c-Jun is required for tumor initiation within 5 weeks after DEN treatment. c-Jun protects hepatocytes from DEN-induced cell death at 48 h after DEN treatment. However, c-Jun has apparently no effect on the activation of the p53 pathway induced by DEN exposure. Interestingly, c-Jun negatively regulates c-Fos mRNA expression at 8 h, but not in liver tumors 8 months after DEN treatment. Moreover, c-Fos alone does not appear to affect liver cancer development, whereas inactivation of both c-Fos/c-Jun restores liver tumor development to levels comparable to controls. These data indicate that while c-Jun antagonizes the pro-apoptotic function of p53 in liver tumors, c-Fos acts as a tumor suppressor at the early stage of tumor initiation in the absence of c-Jun, thus suggesting that c-Jun exerts its oncogenic activity at different levels during liver cancer development.

1. Eferl R, Ricci R, Kenner L, Zenz R, David JP, Rath M, Wagner EF. Liver tumor development: c-Jun antagonizes the proapoptotic activity of p53. *Cell* 2003 112:181–92.

High Throughput 3D Correlative Microscopy

Carlo Tacchetti

IFOM Centre of Cell Oncology and Ultrastructure, Department of Experimental Medicine, University of Genoa, Genova, Italy

Cortese K.*°, Canfora M.*°, Vicidomini G.*°, Gagliani MC*°, Mattioli L.*°, Boccacci P.*°, Diaspro A.*°, Tacchetti C.°*

*MicroscoBio Research Center, University of Genova, and °IFOM, Milano, Italy

By correlative microscopy the same structures observed at the fluorescence light microscopy (FLM) can be analyzed at the EM level. Correlative microscopy resolves several different limitations of fluorescence microscopy. In particular, the absence of a "reference space", e.g., the unlabeled structures not seen in FLM which do not contribute to the analysis. Therefore, it is possible to answer to question as: what is the true size and shape of the structure observed by FLM? However, these studies are extremely time-consuming, especially if statistically relevant data have to be acquired. To overcome these limitations we have set a new correlative method, based on the use semiultrathin cryo-sections, designed to: (a) increase to hundreds the number of events that can be correlated, at each single microscopy session. (b) Obtain 3D correlation between FLM and EM. (c) Provide a free-share software, designed on ImageJ, to perform, in semi- or fully-automatic way, most of the operations needed for the multisection 3D reconstruction of FLM images. (d) Improve the contrast along the Z-axis, at FLM observation, due to the utilization of sections with a subresolution thickness. (e) Allow for further detection of different antigens on the correlated structures at the EM level.

Negative Control of Keratinocyte Differentiation by Rho/CRIK Signaling Coupled with Up-regulation KyoT1/2 (FHL1) Expression

Maddalena Grossi*, Agnès Hiou-Feige*, Alice Tommasi Di Vignano**, Enzo Calautti[†], Paola Ostano[∂], Sam Lee**, Giovanna Chiorino[∂], and G. Paolo Dotto*

*Department of Biochemistry, Lausanne University, Epalinges, Vaud CH-1066, Switzerland
**Cutaneous Biology Research Center, Massachusetts General Hospital and Harvard Medical School, Charlestown, MA 02129, USA and
[∂]Laboratory of Cancer Pharmacogenomics, Fondo Edo Tempia, 13900 Biella, Italy
[†]Epithelial Stem Cell Research Center, The Veneto Eye Bank Foundation, Ospedale Civile di Venezia, Castello 6777, 30122 Venice, Italy

Rho GTPases integrate control of cell structure and adhesion with downstream signaling events. In keratinocytes, RhoA is activated at early times of differentiation and plays an essential function in establishment of cell–cell adhesion. We report here that, surprisingly, Rho signaling suppresses downstream gene expression events associated with differentiation. Similar inhibitory effects are exerted by a specific Rho effector, CRIK (Citron kinase), which is selectively down-modulated with differentiation, thereby allowing the normal process to occur. The suppressing function of Rho/CRIK on differentiation is associated with induction of KyoT1/2, an LIM domain protein gene implicated in integrin-mediated processes and/or Notch signaling. Like activated Rho and CRIK, elevated KyoT1/2 expression suppresses differentiation. Thus, Rho signaling exerts an unexpectedly complex role in keratinocyte differentiation, which is coupled with induction of KyoT1/2, an LIM domain protein gene with a potentially important role in control of cell self-renewal.

Gain of Function Mutant p53: the p53/NF-Y Protein Complex Reveals an Aberrant Transcriptional Mechanism of Cell Cycle Regulation

Silvia Di Agostino[1], Sabrina Strano[1], Velia Emiliozzi[1], Valentina Zerbini[3], Marcella Mottolese[3], Ada Sacchi[1], Giovanni Blandino[1,2], Giulia Piaggio[1,2]

[1] *Department of Experimental Oncology, Istituto Regina Elena, Via delle Messi D'Oro 156, 00158 Rome, Italy*
[2] *Rome Oncogenic Center (ROC), Istituto Regina Elena, Via delle Messi D'Oro 156, 00158 Rome, Italy*
[3] *Department of Pathological Anatomy, Istituto Regina Elena, Via delle Messi D'Oro 156, 00158 Rome, Italy*

Half of human tumors express mutant p53 proteins. Our study demonstrates the oncogenic cooperation of two regulators of the cell cycle, mutant p53 and NF-Y, in proliferating cells and after DNA damage. This cooperation allows cells to override cellular failsafe programs, thus permitting tumor progression. Using cells carrying the endogenous proteins, we demonstrate that mutant forms of p53 protein are engaged in a physical interaction with NF-Y. The expression levels of cyclin A, cyclin B1, cdk1, and cdc25C, as well as the cdk1-associated kinase activities are upregulated after DNA-damage, resulting an increase in DNA synthesis in a mutp53- and NF-Y-dependent manner. Mutp53His175 binds NF-Y target promoters in vivo and, upon DNA damage, recruits p300 leading to histones acetylation. Consistent with this, NF-YA, mutp53His175 and p300 proteins form a triple complex in the cells after DNA damage. In a series of primary human rectal carcinomas with mutant p53, immunoreactivity for NF-Y target genes (*cyclin A* and *cdk1*) was found to be significantly elevated compared with that observed in samples carrying wild-type p53. We provide evidences supporting the hypothesis that mutant p53, through its ability to interact with a variety of transcription factors controls key regulatory activities during the cell cycle.

Stem Cell Epigenetics: Novel Approaches in Conditional Mutagenesis

Giuseppe Testa

European Institute of Oncology, Milan, Italy

Embryonic (ES) and tissue specific (TS) stem cells define the biological problem of how lineage determination and tissue homeostasis are achieved through the coordinated implementation and maintenance of gene expression programs. ES cells constitute a general paradigm of stem cell function and a powerful system to study the balance between self-renewal and differentiation in several lineages, and its disturbance at the level of tumor initiating or cancer stem cells. I focus on the role of histone methylation in the lineage commitment of ES and TS cells. The technology that we recently developed for the fluent engineering of bacterial artificial chromosomes (BACs) through homologous recombination in *E. coli* enables the development of multipurpose alleles, whereby a concerted array of mutations and functional elements are placed into a mouse locus in only one round of gene targeting (Testa et al., Nature 2003, Testa et al., Genesis 2004). Building upon the flexibility of this technology, we are developing novel approaches that combine the ease of BAC engineering with RNA interference with the aim of generating versatile ES and mouse lines in which histone methylation pathways can be dissected with precision through conditional inactivation.

EGFR Overexpression in Malignant Pleural Mesothelioma: Anti-immunohistochemical and Molecular Study with Clinico-Pathological Correlations

Destro A[1], Ceresoli Gl[1], Falleni M[3], Zucali Pa[2], Morenghi E[4], Bianchi P[1], Pellegrini C[3], Cordani N[3], Vaira V[3], Alloisio M[5], Rizzi A[6], Bosari S[3], Roncalli M[7]

[1] Molecular Genetics Laboratory, Istituto Clinico Humanitas, Via Manzoni 56, 20089 Rozzano, Milano, Italy
[2] Department of Oncology, Istituto Clinico Humanitas, Via Manzoni 56, 20089 Rozzano, Milano, Italy
[3] Department of Medicine, Surgery and Dental Sciences, Division of Pathology, University Of Milan, A.O. S. Paolo and Irccs Ospedale Maggiore, Via A. Di Rudin 8, 20142 Milan, Italy
[4] Clinical Trial Office, Istituto Clinico Humanitas, Via Manzoni 56, 20089 Rozzano, Milano, Italy
[5] Department of Thoracic Surgery, Istituto Clinico Humanitas, Via Manzoni 56, 20089 Rozzano, Milano, Italy
[6] Department of Thoracic Surgery, Humanitas Gavazzeni, Via Mauro Gavazzeni, 21 – 24125 Bergamo
[7] Department of Pathology, University of Milan, Istituto Clinico Humanitas, Via Manzoni 56, 20089 Rozzano, Milano, Italy

The epidermal growth factor receptor (egfr) is overexpressed in many epithelial malignancies, against which some antitumoral drugs have been developed. There is a lack of information as to egfr expression in malignant pleural mesothelioma (mpm), an aggressive and fatal cancer poorly responsive to current oncological treatments. Our aim was to (a) compare egfr immunohistochemical expression with mrna levels measured by real time pcr, (b) assess the relationships between egfr expression and clinico-pathological data including survival, (c) analyze the egfr mutations. We developed an immunohistochemical method of egfr evaluation based on the number of immunoreactive cells and staining intensity in 61 mpms. Egfr immunoreactivity was documented in 34/61 (55.7%) cases. a significant correlation between egfr protein and mrna levels ($p = 0.0077$) was found, demonstrating the reliability of our quantification method of egfr membrane expression. Radically resected patients ($p = 0.005$) and those with epithelial histotype ($p = 0.048$) showed an increased survival. No statistical correlation between egfr immunoreactivity and patients' survival was observed. No egfr mutation was documented. This study documents egfr overexpression in mpm at the protein and the transcriptional levels; it proposes a reliable method for egfr expression evaluation in mpm. Egfr levels are not associated with clinico-pathological features of patients, including survival.

hSNM1B is a Novel Telomeric Protein and Functions Together with TRF2 to Protect the Telomeres

Christelle Lenain, Serge Bauwens, Simon Amiard, Marie Josèphe Giraud-Panis, Michele Brunori, and Eric Gilson

Laboratoire de Biologie Moleculaire de la Cellule, UMR5161, Ecole Normale Superieure de Lyon, 46 allee d'Italie, 69364 Lyon Cedex 07, France

Telomeres are essential to protect eucaryotic chromosome ends from degradation and repair. Moreover, modulations of their functionality control cell proliferation and survival. In human cells, the telomeres form large nucleoprotein complexes nucleated around specific telomeric DNA binding factors, namely TRF1, TRF2, and Pot1. Each of these proteins interacts with several additional factors, some of them being involved in other cellular pathways, such as DNA damage response and various types of DNA and chromatin transactions. The precise DNA arrangement at chromosomal ends and the proteins involved in its maintenance are of crucial importance for genome stability. Here, we identify a novel TRF2-interacting factor named hSNM1B, which is a member of the Pso2/Snm1/Artemis b-CASP metallo-b-lactamase family of proteins that function as DNA caretakers. GST-pull down and immunolocalization experiments demonstrate that hSNM1B is associated with TRF2 in vivo and is highly concentrated at telomeres. Knockdown of SNM1B has no detectable effect on chromosomes end protection but increases the telomere dysfunction triggered by the removal of TRF2 from telomeres. We are currently investigating the molecular mechanisms by which hSNM1B cooperates with TRF2 to protects the chromosomal extremities.

Opposite Functions of Dril1, a Member of the ARID Family of DNA-Binding Proteins, in Human and Mouse Fibroblast Senescence

Alexandre Prieur and Daniel S. Peeper

Division of Molecular Genetics, Netherlands Cancer Institute, Amsterdam, The Netherlands

Oncogene-induced senescence (OIS) is triggered when primary cells are exposed to expression of activated oncoproteins (like RASV12), or to hyperactivation of promitogenic proteins resulting from inactivated tumor suppressor genes (like PI3K activity induced by PTEN loss). Recently, we and others have provided the first evidence that OIS represents a genuine physiologically relevant mechanism, protecting mammals against cancer. Previously, we have designed a powerful OIS cell system allowing for a cDNA library-based functional genetic rescue screen, which has identified novel oncogenes, including KLF4 and DRIL1. We are currently characterizing these genes, in particular the mechanism by which they can contribute to oncogenic transformation. Unexpectedly, we found that DRIL1, a member of the ARID family of DNA-binding proteins, has opposite effects in primary human and mouse fibroblasts, in which it induces, or instead bypasses senescence, respectively. We have performed a structure–function analysis to characterize domains within DRIL1 involved in these phenomena. We are also in the process of identifying DRIL1 target genes involved in these phenotypes, by performing a candidate gene approach and microarray analysis. Collectively, these approaches should unravel DRIL1-dependent signaling pathways that contribute to senescence and oncogenic transformation.

Mad2 and the Spindle Assembly Checkpoint: Of Templates, Copies

Andrea Musacchio

European Institute of Oncology, Via Ripamonti 435 I-20141 Milan, Italy

Our laboratory is interested in the molecular bases of formation of stable kinetochore-microtubule attachments during mitotsis. This process is monitored by the spindle assembly checkpoint (SAC), whose components are recruited to kinetochores early in mitosis. Unattached or incorrectly attached kinetochores are the source of a signal that arrests cells in a prometaphase-like state. Metaphase, which entails the biorientation of all sister chromatid pairs, marks the end of the attachment process. This condition switches off the SAC, allowing chromosome separation and anaphase. My laboratory explores the regulation of Mad2, an essential protein component of the SAC. We developed a model for the activation of Mad2 that has the potential to explain the propagation of a cytoplasmic signal from unattached kinetochores. This model entails a self-amplifying positive feedback loop controlling a conformational transition in the Mad2 protein. In many aspects, this mechanism closely resembles the conversion of prion proteins, being based on a template: copy relationship of two Mad2 conformers. I will present recent data from our laboratory obtained in collaboration with S. Piatti (University of Milan-Bicocca) and E.D. Salmon (University of North Carolina, Chapel Hill) showing that the mechanism of Mad2 activation in the SAC is conserved from yeast to humans.

A Membranome to Identify Antibody Candidates with Applications in Oncology

Armin Lahm, Maria Ambosio, Virginia Ammendola, Mirko Arcuri, Alessandro Bellini, Tiziana Brancaccio, Stefano Colloca, Alessandra De Pra, Manuel de Rinaldis, Alessandra Luzzago, Paolo Monaci, Alfredo Nicosia, Laura Orsatti, Fabio Palombo, Claudia Santini, Fabio Talamo, Alessandra Vitelli, and Riccardo Cortese

Istituto di Ricerche di Biologia Molecolare (IRBM) via Pontina, Pomezia, Rome, Italy

Identification of antibody candidates for a potential therapeutic application in oncology normally proceeds through the initial identification of a suitable target antigen and subsequent targeted antibody selection. Here we describe a complementary approach aimed at the generation of a large panel of antibodies against cell surface antigens, with potential application in many therapeutic fields and, in particular, oncology. Starting from a large and complex collection of phage-displayed single-chain antibodies successive rounds of whole cell selection procedures are applied generating a large and diverse collection of antibodies. Implementation of a medium-throughput IgG-conversion procedure supports the generation of large amounts of IgG protein necessary to perform functional characterization of the antibody and also antigen target identification. An overview of our approach will be presented together with examples of antibodies from the collection having antiproliferative in vitro properties.

Comparative Proteomic Analysis of ING1 Protein Interactions in Human Cells

Michael Russell[1], Morgan Hughes[2], Karl Riabowol[1], David Schriemer[1,2]

[1] Department of Biochemistry and Molecular Biology, Faculty of Medicine, University of Calgary, Calgary, AB, Canada
[2] Southern Alberta Mass Spectrometry Faculty, University of Calgary, Calgary, AB, Canada

The ING1 tumor suppressor impinges upon a variety of cellular processes including proliferation, apoptosis, senescence, and the DNA damage response. The function of ING1 in these pathways has largely been deduced through characterization of the proteins that interact with ING1. We set out to uncover novel ING1 protein interactions in human cells with the hopes of further defining the functional role of ING1. ING1 and associated proteins were enriched from HEK293 cells using a panel of ING1 monoclonal antibodies and were identified using one of three approaches: (1) direct liquid chromatography–tandem mass spectrometry (LC–MS/MS) analysis; (2) prefractionation by 1D SDS–PAGE followed by LC–MS/MS analysis or; (3) a MudPIT-like method. Several novel ING1-binding partners were identified, including chromatin-associated proteins, mRNA splicing proteins, and stress response proteins, all consistent with known and/or speculated functions for ING1. Putative ING1-interacting proteins were subdivided based upon biological function and molecular pathway using the PANTHER algorithm with the hope of placing ING1 in a broad functional context with respect to its binding partners and the pathways in which it is involved. In addition, the three experimental approaches were compared with respect to protein interaction discovery capability in an endogenous expression system.

Identification of Novel Tumor Suppressor Genes in the p16INK4a/pRB and p53 Pathways

Moroni Mc[1], Sulli G[1], Bettolini R[1], and Helin K[1,2]

[1] IFOM-IEO-Campus, Milano, Italy
[2] BRIC, Copenhagen, Denmark

Cancer is a heterogeneous disease, which arises from the accumulation of multiple genetic and epigenetic alterations, which lead to the activation of oncogenes or the inactivation of tumor suppressor genes. Normal human cells from different origin can become transformed by the introduction of at least five genetic alterations, including: hTERT expression, the inhibition of both the p53 and the p16INK4a/pRB pathways, the overexpression of the H – RASV12 oncogene and the expression of the SV40 small t antigen. In order to identify tumor suppressors which can functionally substitute, respectively, for the inactivation of pRB or p53 in this in vitro model, we are performing two kinds of loss of function screens. We generated two different pools of transformation – prone human fibroblasts, which contain all the above-mentioned alterations, but lack the inactivation of either the p53 or the p16INK4a/pRB pathway. The omitted genetic alteration has been substituted by the introduction of a retroviral shRNA library, followed by selection of the cells which become transformed, i.e., able to grow as anchorage independent colonies. We identified several transforming shRNAs in these screens. We are currently validating and prioritizing the corresponding candidate tumor suppressors, for initial characterization.

Genetic and Epigenetic Changes in the Progression of Colorectal Cancer

Miranda E[1], Bianchi P[1], Balladore E[1], Destro A[1], Franchi G[1], Baryshnikova E[1], Morenghi E[2], Laghi L[1-3], Gennari L[4] Malesci A[5], Roncalli M[1,6]

[1] *Molecular Genetics Laboratory, Istituto Clinico Humanitas, Via Manzoni 56, 20089 Rozzano, Milano, Italy*
[2] *Clinical Trial Office, Istituto Clinico Humanitas, Via Manzoni 56, 20089 Rozzano (Milano), Italy*
[3] *Departement of Gastroenterology, Istituto Clinico Humanitas, Via Manzoni 56, 20089 Rozzano (Milano), Italy*
[4] *Department of Surgery, Istituto Clinico Humanitas, Via Manzoni 56, 20089, Rozzano (Milano), Italy*
[5] *Departement of Gastroenterology, Istituto Clinico Humanitas, University of Milan, Via Manzoni 56, 20089, Rozzano (Milano), Italy*
[6] *Departement of Pathology, Istituto Clinico Humanitas, University of Milan, Via Manzoni 56, 20089 Rozzano, (Milano), Italy*

CRC develops as multistep process which involves genetic and epigenetic alterations. K-Ras, p53 and B-Raf mutations and RASSF1A, E-Cadherin and p16INK4A promoter methylation were investigated in 206 primary colorectal cancers to assess whether gene abnormalities are related to tumor progression. K-Ras, B-Raf and p53 mutations were detected in 27%, 3%, and 33% CRCs. K-Ras mutations were significantly associated with advanced Astler&Coller C2 and D tumor stage ($p = 0.013$) and with the right side ($p = 0.028$). RASSF1A, E-Cadherin and p16INK4A promoter methylation was documented in 20%, 43%, and 33% with p16INK4A significantly associated with advanced tumor stage ($p = 0.0001$) and E-Cadherin with the right side ($p = 0.02$). Overall, 34/206 tumors (16%) did not show any molecular change, 129/206 (63%) had 1 or 2 and 43 (21%) 3 or more. Metastatic tumors were prevalent in the latter group ($p = 0.0001$) where the most frequent combination was one genetic and two epigenetic alterations. This analysis provided to detect some molecular differences between primary metastatic and nonmetastatic CRCs. K-Ras and p16INK4A were altered during progression, while specific gene combinations, such as coincidental K-Ras mutation and two methylated genes, were mainly associated with a metastogenic phenotype.

Dominant Negative Mutation of the Fumarate Hydratase Tumor Suppressor Gene

Annalisa Lorenzato[1], Martina Olivero[1], Mario Perro[1], Emmanuel Dassa[2], Pierre Rustin[2], and Maria Flavia Di Renzo[1]

[1] *Laboratory of Cancer Genetics, Institute for Cancer Research and Treatment (IRCC), University of Turin Medical School, Str. Provinciale 142, Km 3.95, 10060 Candiolo, Turin, Italy*
[2] *INSERM U676, Hospital Robert Debrè, 48 Boulevard Serurier, 75019 Paris, France.*

Germline heterozygous mutations in *fumarate hydratase* (FH) gene predispose to hereditary leiomyomatosis, which is associated in 30% of cases to renal-cell carcinoma (HLRCC), showing a highly aggressive phenotype. The FH encodes a mitochondrial tumor suppressor, showing the link between mitochondria dysfunction and tumorigenesis. FH functions as a homotetramer and a dominant-negative effect of some FH missense mutations has been proposed. In this work we analyzed the I77V, R190H and E319Q FH mutations in cell models. We first expressed wild type and mutated proteins in FH−/− skin fibroblasts, which are naturally occurring fumarase deficient cells. Cells overexpressing wild-type and I77V *FH* transgenes were able to restore the fumarase activity while the overexpression of the R190H and E319Q FH did not increase the enzyme's activity. More interestingly, when the same transgenes where expressed in cells containing endogenous wild-type fumarase, the R190H mutated fumarase reduced the endogenous FH activity up to 70%. These data demonstrate for the first time the dominant negative effect of an *FH* gene mutation and suggests that mutations might have a different functional outcome and possibly phenotypic consequences.

Mutant P53 Gain of Function: Reduction of Tumor Malignancy of Human Cancer Cell Lines Through Abrogation of Mutant P53 Expression

Gianluca Bossi, Eleonora Lapi, Sabrina Strano, Cinzia Rinaldo, Giovanni Blandino, and Ada Sacchi

Department of Experimental Oncology, Regina Elena Cancer Institute, Rome, Italy

Mutations in the TP53 tumor suppressor gene are the most frequent genetic alteration in human cancers. These alterations are mostly missense point mutations that cluster in the DNA binding domain. There is growing evidence that many of these mutations generate mutant-p53 proteins that have acquired new biochemical and biological properties. Through this gain of function activity mutant-p53 is believed to contribute to tumor malignancy. The purpose of our study was to explore mutant-p53 as a target for novel anticancer treatments. To this aim, we inhibited mutant-p53 expression by RNA interference in three different cancer cell lines endogenously expressing mutant-p53 proteins, and evaluated the effects on the biological activities through which mutant-p53 exerts gain of function. We found that, depletion of mutant-p53 reduces cell proliferation, in vitro and in vivo tumorigenicity, and resistance to anticancer drugs. Our results demonstrate that mutant-p53 knockingdown weakens the aggressiveness of human cancer cells, and provides further insides for the comprehension of mutant-p53 gain of function activity in human tumor. Studies are now ongoing to identify putative mutant-p53 target genes through/with which mutant-p53 exerts gain of function activity.

P53 Functional Down-Regulation by NPM-ALK and other Hematopoietic Oncoproteins

Paola Martinelli, Emanuela Colombo, Pier Giuseppe Pelicci.

European Institute of Oncology, via Ripamonti 435, Milan, Italy

p53 mutations are rare in hematopoietic malignancies, as opposed to other types of tumors. This raises the idea that additional mechanisms of p53 inactivation may be involved. PML-RAR and Bcr–Abl are two examples of p53 functional inactivation by hematopoietic oncogenes described so far. We tested a number of oncogenic Tyr-kinase fusion proteins, isolated from leukemias and lymphomas. All of them repress p53 activity, suggesting that its functional down-regulation is a key event in the transforming activity of these proteins. Further analysis of the relationship between structure and function of the chimeric protein NPM–ALK revealed that an intact Tyr-kinase activity is required for p53 inhibition, while the presence of NPM portion is not. Surprisingly, despite its repressing action on p53 function, NPM–ALK overexpression in primary MEFs induces a senescence-like growth arrest. Thus, NPM–ALK triggers a cellular checkpoint response, which is p53-independent. pRb has been described as a checkpoint protein, so we investigated the pRb/p16 pathway as a target of NPM–ALK. p16INK4a is activated in wild-type primary MEFs expressing the fusion protein, and loss of either pRb is sufficient for primary cells to overcome NPM–ALK-induced growth arrest. Our data strongly indicate that NPM–ALK expression triggers a p53-independent, pRb/p16-dependent checkpoint response.

A Novel Transgenic Approach to Explore Myc Physiological Functions

Laura Soucek*^, Jonathan Whitfield*, Andrew Finch*, Mauro Savino^, Sergio Nasi^, Gerard I. Evan*

*Comprehensive Cancer Center, UCSF, S. Francisco, USA
^IBPM CNR, Dipatmento. Genetica e Biologia Molecolare, Università La Sapienza, Roma, Italy

c-Myc is well known for its involvement in cancer pathology. However, its physiological function is still not clear. Due to embryonic lethality caused by Myc loss, studies aiming to unveil its function in adult mice have so far relied on conditional knock-outs, which have inherent limitations. Here we employ a novel approach, based on the conditional and reversible expression of a Myc mutant, Omomyc, in adult mice tissues. We show that Omomyc affects the transcriptional program driven by Myc under physiological proliferative conditions, inhibiting its ability to transactivate genes, but facilitating its repressor function. Abrogation of the endogenous Myc transactivation capacity through a tetracycline–responsive Omomyc transgene impairs liver regeneration, but also reveals the ability of Myc to regulate cell cycle duration in the intestinal epithelium, so that prolonged interference with Myc function causes a dramatic shortening of intestinal villi due to a reduced replacement of epithelial cells. Our data are consistent with the hypothesis that Myc has a role in different situations in which high proliferation is required, either because of continuous cellular turn-over or during regeneration of damaged tissues.

Characterization of a Conserved Retinoblastoma Repressor Complex

Michael Korenjak, Barbie Taylor-Harding*, John Satterlee*, Nick Dyson*, and Alexander Brehm

Adolf-Butenandt-Institut, Molekularbiologie, Ludwig-Maximilians-Universitaet Muenchen, Germany
*Massachusetts General Hospital Cancer Center, Charlestown, USA

The role of the Retinoblastoma tumor suppressor (pRb) and other members of the pocket protein family in cell cycle progression has been characterized extensively. However, the molecular mechanisms that pocket proteins use to regulate transcriptional programs during differentiation are poorly understood. Recently, we have purified and characterized the first native pocket protein complex from *Drosophila melanogaster*. This complex localizes to transcriptionally silent chromatin in vivo and represses a set of differentiation specific genes. It contains *Drosophila* RBF, E2F, and Myb-interacting proteins (DREAM) and appears to be conserved in *C. elegans*: the worm homologues of DREAM subunits act in a common genetic pathway which controls the expression of genes important for vulval differentiation. Furthermore, several of the *C. elegans* proteins have been shown to interact with each other. Interestingly, the human genome encodes homologues of all DREAM subunits. Here, we present evidence that these proteins are expressed and interact in human cells. Furthermore, they exist in a high molecular weight complex in several different cell lines. Purification of this complex provides evidence that a complex similar to DREAM operates in human cells. Our results demonstrate a remarkable degree of conservation of pRb complexes implicated in the regulation of transcriptional programs during differentiation.

A Role for Histone Deaceylase 1 in Human Tumor Cell Proliferation

Silvia Senese, Katrin Zaragoza-Dorr, Simone Minardi, Loris Bernard, Giulio F. Draetta, Myriam Alcalay, Christian Seiser, and Susanna Chiocca

Post-translational modifications of core histones are central to the regulation of gene expression. Histone deacetylases (HDACs) repress transcription by deacetylating histones and Class I HDACs have a crucial role in mouse, Xenopus, zebrafish and *C. elegans* development. The role of individual Class I histone deacetylases (HDACs) in the regulation of cell proliferation was investigated using RNAi-mediated protein knockdown. We show here that ablating HDAC1 and HDAC3 protein expression resulted in the inhibition of U2OS cell proliferation and an increase in the percentage of apoptotic cells. On the contrary HDAC2 knockdown showed no effect, unless we concurrently knocked down HDAC1. In line with this finding, RNAi against HDAC1 alone or in combination with HDAC2 increased significantly the expression of p21 protein, a cyclin-dependent kinase inhibitor. We also observed that only when both HDAC1 and HDAC2 were downregulated, histones H3 and H4 became hyperacetylated. Using gene expression profiling analysis, we found that while HDAC2 downregulation did not determine significant changes compared to control cells, inactivation of either HDAC1, HDAC1+2, or HDAC3 resulted in more distinct clusters. In addition, HDAC1 knockdown cells were unable to proceed through mitosis. Our results demonstrate that HDAC1 and HDAC3 play a major role in proliferation and survival of these cells. Loss of these HDACs might impair cell cycle progression not only by affecting the transcription of specific target genes (p21) but also through effects on other important biological processes. Drug targeting specific HDACs could be highly beneficial in the treatment of cancer.

Transcriptional Targeting of Tumor Microenvironment by the Grp94 Stress-Promoter

Isabella Alessandrini, Claudia Chiodoni, Mariella Parenza, Mario P. Colombo

Experimental Oncology-Immunotherapy and Gene Therapy Unit, Istituto Nazionale Tumori, Milan, Italy

Tumor microenvironment is a potent activator of glucose-regulated protein 94 (grp94), the most abundant stress response glycoprotein within the endoplasmic reticulum. Its promoter can drive specific expression of therapeutic genes in the context of solid tumors. The cytokine IL-12 induces tumor regression in mice while it is toxic in human if given systemically. The goal is to express IL-12 in stressed tissues, such as tumors, via the grp94 promoter and bone marrow (BM) cells as vehicle to reach the tumor. Grp94 promoter has been cloned into a third generation self-inactivating lentiviral vector. Preliminary in vitro experiments have shown the induction of IL-12 expression under hypoxia condition. The efficacy of this approach will be tested against different mammary tumor models. The transplantable tumor cell lines N2C (derived from HER-2/neu transgenic mice) and 4T1 that spontaneously metastasize to distant organs. Finally, to mimic a clinical setting of auto-stem cell transplantation contaminated by lymphoma cells, we will use the A20 murine lymphoma mixed to donor "therapeutically-infected" BM cells at titrated minimal doses. We are confident that IL-12 if properly targeted at the tumor site will be able to induce consistent tumor rejection.

Dynamics of Replication Compartments in Response to DNA Damaging Agents

Samuela Soza, Riccardo Vago, Rossella Rossi, Giuseppe Biamonti, Alessandra Montecucco

Istituto di Genetica Molecolare del CNR, via Abbiategrasso 207, Pavia, Italy

In mammalian cells DNA replication takes place in functional subnuclear compartments called replication factories, where replicative factors accumulate. Replication factories are composed of coordinately replicated chromosomal domains and their distribution reflects the replication of different portions of the genome identifying the different moments of the S phase (from early to late). This dynamic organization is affected by agents that induce cell cycle checkpoints activation via DNA damage or stalling of replication forks. We have investigated the cell response to etoposide, an anticancer drugs belonging to the topoisomerase II poisons. We have shown that etoposide does not induce an immediate block of DNA synthesis and progressively affects the distribution of replication proteins in S phase. Moreover it triggers the formation of large nuclear foci that contain the single-strand DNA binding protein replication protein A suggesting that lesions produced by the drug are processed into extended single-stranded regions. The dispersal of replicative proteins, PCNA and DNA ligase I, from replication factories requires the activity of the ataxia telangiectasia Rad3-related (ATR) checkpoint kinase. By comparing the effect of the drug in cell lines defective in different DNA repair and checkpoint pathways we have dissected the etoposide response identifying the proteins involved.

The Set Domain-Negative Form of PRDM16, sPRDM16, Induces Acute Myeloid Leukaemia in Mice

Danielle C. Shing[1], Maurizio Trubia[2], Francesco Marchesi[3], Enrico Rada[3], Eugenio Scanziani[3], Massimo Stendardo[2], Silvia Monestiroli[2], and Pier Giuseppe Pelicci[1]

[1] *Department of Experimental Oncology, European Institute of Oncology, Milan, Italy*
[2] *IFOM-FIRC Institute of Molecular Oncology, Milan, Italy*
[3] *Department of Veterinary Pathology, Hygiene and Public Health, School of Veterinary Medicine, University of Milan, Italy*

PRDM16 at 1p36 is rearranged in acute myeloid leukaemia (AML). The PRDM16 protein exists as two isoforms that differ in the presence and absence of the SET domain. The SET domain-negative form, sPRDM16, is aberrantly overexpressed in t(1;3)(p36;q21)-positive AML and in some lymphoid leukemias. We have investigated the oncogenicity of SET domain-positive (PRDM16) and SET domain-negative (sPRDM16) isoforms. In vitro, expression of sPRDM16 in murine bone marrow progenitor cells blocks myeloid differentiation and increases in self-renewal. In vivo, mice transplanted with bone marrow progenitor cells expressing sPRDM16 develop AML. If donor cells are wild-type (p53+/+), sPRDM16 induces AML within 5–6 months in 40% of the mice (experiment ongoing). However, sPRDM16 induces AML with a shorter latency of 2–5 months and with 100% penetrance (13 mice) when expressed in p53–/– donor cells. Mice transduced with empty vector or PRDM16 do not develop leukaemia. The myeloid leukemias induced by sPRDM16 in the absence of p53 are characterized by erythroid and dysplastic megakaryocyte components and are transplantable into secondary recipient mice. Therefore, the mouse model we have created reflects the human disease and will serve as a valuable tool for studies on AML. In addition, we demonstrate that PRDM16 is oncogenic only in the absence of its SET domain.

Molecular Alterations in Sputum of Heavy Smokers at High Risk of Developing Lung Cancer

Dr. Ekaterina Baryshnikova, Dr. Annarita Destro, Dr. Maurizio Infante, Prof. Massimo Roncalli

Istituto Clinico Humanitas - Irccs, Via Manzoni 56, 20089 Rozzano (Mi), Italy

Lung cancer, the leading cause of tumor-related death, is a key example of a cancer for which mortality could be greatly reduced through the development of sensitive molecular assay to be used in screening of high-risk subjects, like heavy smokers. the prevalence of specific molecular markers could approximate life time risk for developing lung cancer. In our study, we analyzed the frequency of point mutations of k-ras and p53 and promoter methylation of p16, rassf1a, and nore1a genes in formalin-fixed paraffin-embedded samples of sputum in male current and former heavy smokers older than 60 years. We found no mutation of k-ras, supporting the idea that it is characteristic of developed cancer and thus unuseful in early diagnostic. almost one alteration in *p53* and *p16* genes was detected in 6.5% of analyzed subjects (50 of 773) in spite of no clinical and radiological evidence of cancer. These patients have been enrolled in a follow-up study and one of them developed lung cancer after 2 years. So, *p16* and *p53* are more suitable to be adopted in screening of early stage lung cancer.

Differences in the Transcriptional Regulation of PDGF Receptor Beta in Neuroblastoma

Daniel Wetterskog, Weiwen Yang, and Keiko Funa

Institute of biomedicine, Goeteborg University, Gothenburg, Sweden

PDGF receptor beta (PDGFRB) is known to be conditionally regulated in the nervous system, especially in late development and in the adult. It is also highly expressed in some types of cancer. We therefore raised the question if neuroblastoma, being derived from neural crest cells, had abnormal regulation of PDGFRB expression. Expression levels of PDGFRB protein and mRNA decreased after serum stimulation in SH-SY5Y neuroblastoma cells similar to what we previously reported in NIH3T3. However, the IMR32 neuroblastoma showed stable levels of PDGFRB after serum stimulation. Using flow cytometry, we followed cell cycle progression in the cell lines upon serum re-addition. The two cell lines showed distinctly different profiles in their cell cycle progression. However, when the cell lines were treated with the DNA damage-inducing agent cisplatin they displayed similar cell cycle profiles and the expression levels of PDGFRB decreased in both cell lines. Since PDGFRB expression is known to be regulated by p53 family members we did chromatin immunoprecipitation with antibodies against p53 family members at different periods after serum stimulation or treatment with cisplatin. Our results indicate that p53, directly interacts with the PDGFRB promoter, regulating transcription of PDGFRB. Differences between the two cell lines will be discussed.

Chromatin Mediated Regulation of V(D)J Recombination

Grazini U., McBlane F.

European Institute of Oncology, Milan, Italy

Chromosomal V(D)J recombination generates antigen specific B and T cell receptors during lymphocyte development. The imprecise joining of multiple genomic segments generates functional antigen receptor genes. DNA recombination activity is targeted by conserved recombination signal sequences (RSS), which are recognized by a complex containing two lymphoid-specific proteins, RAG1 and RAG2. RAG expression and accessibility to the genome need to be tightly regulated: inefficient recombination results in immunodeficiency, while inaccuracies can result in oncogenic chromosome rearrangements. RAG1 contains the catalytic domain involved in the generation of DNA breaks within Immunoglobulin and *TCR* genes but how such activity is regulated in vivo is poorly understood. Using reconstituted chromatin templates and purified RAG proteins, we developed an in vitro approach to investigate the role of chromatin remodelers in modulating RAGs accessibility to RSS. Our results show that RAGs themselves act as co-factors in regulating accessibility to RSS in chromatin. Mutations or deletions of RAG1 N-terminus or RAG2 C-terminus reduce their accessibility to RSS on chromatinized templates, suggesting they could directly interact with chromatin or chromatin remodeling proteins. Moreover, we have identified novel targets for the E3 ubiquitin ligase activity of RAG1, strengthening the prediction of RAG1's active role in target site selection.

Chemoresistance of Tumor Cells Linked to Mutant P53/P73 Protein Complex

Olimpia Monti, Silvia Di Agostino, Ada Sacchi, Sabrina Strano, and Giovanni Blandino

Department of Experimentally Oncology, Regina Elena Cancer Institute, 00158-Rome, Italy

Cancer might result from both the aberrant activation of oncogenes and the inactivation of tumor suppressor genes. Among the latter, p53 is a considered a key tumor suppressor gene that is inactivated by missense mutations in half of human cancers. Several in vitro and in vivo evidences have shown that the resulting mutant p53 proteins gain oncogenic properties favoring the insurgence, the maintenance, the spreading of malignant tumors and the resistance to conventional anticancer treatments. One potential molecular mechanism underlying gain of function of mutant p53 consists of its ability to physically interact, sequester and inactivate tumor suppressor proteins: The formation of these complexes results in the severe impairment of p73-mediated transcriptional activity and apoptosis. To establish a critical role of the protein complex mutant p53/p73 in the response of tumor cells to anti-cancer treatment, we have engineered short synthetic peptides capable to disassemble that protein complex. Treatment of tumor cells bearing mutant p53 with the short interfering peptides enhances cisplatin and doxorubicin-induced apoptotic response that occurs through the transcriptional activation of apoptotic p73 target genes. Our findings identify the protein complex mutant p53/p73 as a key target whose successfully overriding might improve cancer therapy.

Fra-1 and Fra-2 are Antagonists of p53

Sara Piccinin*, Alessandra Grizzo*, Silvia Demontis*, Maura Sonego*, Michela Armellin*, Marika Garziera*, Greg J. Hannon[§], and Roberta Maestro*

*Experimental Oncology 1, CRO National Cancer Institute, 33081 Aviano, Italy
[§] Cold Spring Harbor Laboratory, 11724 Cold Spring Harbor NY, USA

Fra-1 and Fra-2 (Fos-related antigen 1 and 2) are nuclear proteins of the Fos family of transcription factors that, together with jun proteins, participate in the formation of the AP-1 complex. Although overexpression of Fra-1 and Fra-2 has been reported in a number of human tumor types, their role in cancer development has only been marginally clarified. Here we report that Fra-1 and Fra-2 are potent antagonists of the p53 pathway. Ectopic expression of Fra-2, and in part Fra-1, render cells resistant to p53-mediated apoptosis and growth arrest. This antagonism toward p53 is associated to a reduced induction of p53 following oncogenic insults. We provide evidence that Fra proteins enhance p53 degradation by promoting its ubiquitination. We are currently investigating the role of MDM2 in Fra-induced p53 degradation. Our data support a role for Fra-1 and Fra-2 as important modulators of p53 and suggest a crosstalk between AP-1 and p53 during apoptosis and tumor development.

In Vivo Epigenetic Pharmacotherapy Against Leukemias

Soncini Matias, Saverio Minucci

European Institute of Oncology, Milan, Italy

The histone deacetilase inhibitor valproic acid (VPA) demonstrated to contrast leukemia progression in PML/RAR leukemic mice by inducing apoptosis via FAS and TRAIL pathways selectively on leukemic blasts. We have demonstrated that another epigenetic drug, 5 aza 2'deoxycytidine (DAC), which targets DNA methyl-transferases, significantly increases leukemic mice survival in a comparable way to VPA. It induces in vivo apoptosis of leukemic blast and expression of the same death receptors pathways of VPA, and provokes blast depletion in bone marrow and peripheral blood. TRAIL fundamental role has been confirmed by in vivo siRNA. In vitro DAC demonstrated to be less selective than VPA on leukemic blasts, but in vivo we didn't noticed any strong negative effect on mice. The combination of the two drugs synergizes both in vitro and in vivo further increasing leukemic mice survival. DAC and VPA both improve the all *trans*-retinoic acid (ATRA, a differentiating drug) antileukemic effect. Combination of DAC, VPA, and ATRA is really effective, but showed lethal myelosuppression in 66% of treated leukemic mice and in 50% of healthy mice.

P63 Protects the Female Germline by Eliminating Damaged Oocytes During Meiotic Arrest

Eun-Kyung Suh, Annie Yang, Ala H. Michaelis, Casimir Bamberger, Julia A. Elvin, Christopher P. Crum, and Frank Mckeon

Department of Cell Biology, Harvard Medical School, Havard, USA

Meiosis in the female germline of mammals is marked by a prolonged oocyte arrest in prophase of meiosis I prior to ovulation. During this long period of arrest, these post-recombination, primary oocytes are vulnerable to genomic damage. How DNA damage is detected in oocytes is poorly understood but variably thought to involve p53. Here we show that p63, and not p53, plays an essential role in oocyte death following DNA damage. Our data support a model whereby p63 acts in a conserved process of monitoring the integrity of the female germline, while p53 functions are restricted to vertebrate somatic cells for tumor suppression. These findings have implications for understanding the maintenance of germline fidelity, the significance of female meiosis therein, and the evolution of tumor suppressor mechanisms.

Transient and Reversible Epithelial to Ameboid Transition in Metastatic Breast Cancer Cells

Silvia Giampieri and Erik Sahai

Tumor Cell Biology Laboratory, CRUK London Research Institute

Metastasis is a multistep process that requires tumor cells to become motile, enter, and exit the vasculature and successfully colonise a target organ. In order to directly study this process we have used multiphoton intravital imaging of a metastatic breast tumor model. By fluorescently labeling the cytoskeleton of the tumor cells, we monitored the behavior of cells within the primary tumor and disseminated cells in the lymph nodes of living mice. This has lead to the following observations: (1) The majority of tumor cells in metastatic primary tumor are nonmotile and have an "epithelial" organization. (2) Approximately 1% of cells in metastatic tumors are highly motile moving at speeds of up to 15 $\mu mmin^{-1}$. These cells are localized to specific microenvironments within the tumor and have an "amoeboid" morphology, lacking distinct focal adhesions. (3) Cells preferentially metastasize to a peripheral region of local lymph nodes, where they cease to be motile and revert to a more epithelial morphology. These observations help to account for the frequent but puzzling phenomenon that metastases of human breast cancers can have an epithelial appearance and suggest that transient and reversible changes in tumor cell behavior are crucial for the metastatic process.

Translational Pharmacogenomics to Tailor Chemotherapy for Nonsmall Cell Lung Cancer

Francesca Toffalorio[1], Romano Danesi[2], De Braud[3]

[1] IEO, Milan
[2] University of Pisa, Pisa, Filippo
[3] IEO, Milan

Lung cancer is a leading cause of mortality in the Western world. Most patients' present advanced disease at diagnosis and chemotherapy treatment remains the standard approach. Unfortunately, the efficacy of chemotherapy is still poor, with less than 25% response rate in patients with advanced nonsmall cell lung cancer (NSCLC) treated with a single agent. Most antineoplastic compounds undergo cellular bioactivation to become pharmacologically active. Several enzymatic activities, involved in their biotransformation, show a significant interindividual variability, leading to changes in efficacy and tolerability to the chemotherapeutic agent. Pharmacogenomics is the science that examines the genetic basis of variation in drug metabolism, drug targets, and transporters, providing the molecular basis for rationally choosing the correct drug for the patient. We have previously established an approach to define the NSCLC pharmacogenomic profile by directly studying microdissected tumor samples by quantitative PCR and DNA polymorphism analysis. In particular, we have focused our attention on the following genes involved in drugs activity: deoxycytidine kinase (dCK), deoxycytidine deaminase (CdA), endo-5'-nucleotidase (5'-NT), hENT and ribonucleotide reductase (RR) for gemcitabine, ERCC1 for cisplatin, topoisomerase I and UGT1A for irinotecan, beta-tubulin for docetaxel or paclitaxel. Our current study consists of patients with advanced NSCLC that undergo surgical biopsy and then "empiric" standard chemotherapy, irrespective of the pharmacogenomic profile. In these patients, we identify and characterize gene expression profiles and threshold of gene products that could be related to chemosensitivity and/or resistance to a given single drug. Instead, in the second part, treatments will be tailored according to our "therapeutic efficacy assessment" related to the pharmacogenomic profile in order to maximize response to treatment and to improve overall survival. The aim of the project is to assess the criteria for matching chemotherapy with the biological features of lung cancer in order to: (1) overcome the limitation of an empiric choice of the drug treatment administered (2) identify the relationship between gene products that serve as predictive factors of drug activity and predictive factors of tumor response (3) develop methods (i.e., monoclonal antibodies, standard PCR) as tools to allow for such analysis to be performed on a routine basis in the clinical setting.

Role of RaLP, a Novel SHC Family Member, in Melanoma Progression

Ernesta Fagiani, Giuseppina Giardina, Giovanni Germano, Micaela Quarto, Maria Capra, Lucilla Luzi, and Luisa Lanfrancone

Department of Experimental Oncology, European Institute of Oncology, Via Ripamonti 435, 20141 Milano, Italy

RaLP is a recently isolated member of the Shc family of adaptor proteins, that shares the same modular organization (CH2, PTB, CH1, and SH2 domains). We have demonstrated that it is a physiological substrate of various receptor tyrosine kinases. Analysis of RaLP expression in normal adult tissues and in a panel of 350 primary tumors of different histological origin showed selected expression in melanomas. Moreover, during melanoma progression RaLP is expressed in vertically growing and metastatic melanomas, but not in nevi and radially growing melanomas, thus suggesting a role of RaLP in the invasive and migratory phenotype of this tumor. Down-regulation of RaLP expression in melanoma inhibits cell migration and induces apoptosis. Primary cultures of normal melanocytes are not transformed by RaLP overexpression. Together, these data suggest that increased expression of RaLP protein in metastatic melanomas is essential, but not sufficient, to maintain a transformed phenotype. Analysis of RaLP expression during embryonic development reveals high levels of RaLP protein product in neural stem cells and migrating melanoblasts, which decreases in the differentiated melanocytes. We propose that RaLP is critical for the maintenance of the neuronal/melanoblast stem cell phenotype during embryo development, and that this function might contribute to tumorigenesis when RaLP is overexpressed in adult cells of melanocytic origin.

Targeting the Cell Cycle and the P13K Pathway: A Universal Strategy to Reactivate Innate Tumor Suppressor Programs in Cancer Cells?

Therese David-Pfeuty[1], Michel Legraverend[2], Odile Ludwig[2], and David S. Grierson[2]

Institut Curie-Recherche, [1]UMR 146 and [2]UMR 176 du CNRS, Batiment 110, Centre Universitaire, 91405 Orsay Cedex, France

Corruption of the Rb and p53 modules provides a surfeit of Cdk activity and growth enabling oncogene-bearing cells to elaborate strategies to evade tumor-suppressive mechanisms. A universal means to palliate their deficiency in cancer cells might be therefore to target both the activity of Cdks and their PI3K module, that regulates cell growth. Consistently, we observed that the potentiation of roscovitine-induced apoptosis by the PI3K inhibitor, LY294002, and the killing efficacy of 16 purines decreased with increasing corruption of the Rb and p53 modules. Furthermore, we found that purines differing by a single substitution, which exerted little lethal effects on distant cell types grown on a rich medium, could display widely-differing cytotoxicity profiles toward the same cell types grown on a poor medium. This highlights that structurally related compounds may target, beside the same Cdks, unique proteins or Cdks controlling unique biological functions or competing for their interaction with therapeutically relevant Cdk targets. Therefore, in the perspective of clinical development, tumor cell type-selective Cdk inhibitors should be selected according to their toxicity in cell-based assays performed at a limiting serum concentration, sufficient to suppress their interaction with undesirable crossreacting targets (whose range and concentration would depend on the cell genotype).

A Role for RNA Interference in Polycomb-Mediated Transcriptional Silencing, Independent of Chromatin

Filippo M. Cernilogar[1], Achim Breiling[1], Axel Imhof[2], and Valerio Orlando[1]

[1] *Dulbecco Telethon Institute, Institute of Genetics and Biophysics IGB CNR, Via Pietro Castellino 111, 80131 Naples, Italy*
[2] *Adolf-Butenandt Institut University of Munich Schillerstr. 44 80336 Munich, Germany*

The *polycomb group* (*PcG*) genes code for components of multiprotein complexes that control epigenetic inheritance of gene silencing of tissue specific- and developmentally-regulated genes. The mechanisms underlying PcG repression are rather complex. These appear to involve (i) post-translational modification of histone octamer (ii) compaction of chromatin fiber (iii) constitutive block of RNA Polymerase and (iv) subnuclear compartmentalization. Recent breakthroughs in gene regulation revealed an unforeseen link between synthesis and degradation of noncoding RNA and control of gene expression acting at the level of chromatin structure. We use the fruit fly *Drosophila melanogaster* to investigate the role of PcG, RNAi, noncoding RNA in epigenome reprogramming. Data will be presented in support of a novel, direct involvement of RNAi components in transcriptional gene silencing acting in cooperation with PcG cell memory system in the epigenetic control of homeotic genes, independent of chromatin.

The Role of SUMO in the Stability Diverse Transcriptional Complexes

Paula Babarovic and Susanna Chiocca

Department of Experimental Oncology, European Institute of Oncology 20141 Milan, Italy

Post-translational modifications of proteins have critical roles in many cellular processes as they can cause rapid changes in the functions of preexisting proteins, multiprotein complexes and subcellular structures. One of recently discovered post-translational modification system is sumoylation. This modification involves four steps and different enzymes: SUMO activating enzyme E1, SUMO conjugating enzyme E2 (UBC9), SUMO ligases E3s and SUMO specific proteases. SUMO is emerging as a modifier for a large number of proteins in many different pathways with diverse consequences. Available data currently implicate SUMO in the regulation of protein–protein interaction, subcellular localization, protein–DNA interactions and, in some cases, as an antagonist of ubiquitination. One of the proteins known to be sumoylated is HDAC1 (Colombo et al., 2002), an important component of a variety of transcriptional repressor complexes; such as Sin3A/HDAC, REST/CoREST and NuRD complexes. Although findings have established the importance of HDAC recruitment to DNA through specific DNA-binding complexes, little is understood on how the assembly or deacetylase activity of these complexes is regulated. Until now phosphorylation of specific serine residues (S421, S423) has been implicated in regulation of the complex assembly and activity. We would like to see what is the effect of sumoylation in the same complex formation and activity. Preliminary results obtained in our lab are indicating that also other protein subunits of the Sin3A-HDAC transcription repressor complex are sumoylated. Furthermore, to understand the role of sumoylation on HDAC1, we are using the fusion proteins approach. We obtained HDAC1wt, HDAC1 mutated in targeted SUMO sites (K444, K476R) and HDAC1 mutated in two phosphorylation sites (S421, S423A) fused in frame with SUMO-1. Our preliminary results are pointing to a model in which in the absence of a functional SUMO pathway, the proteins involved in Sin3A/HDAC, REST/CoREST complex formation are impaired to bind to the other components of the complex and are more accessible to the degradation machinery.

Characterization of P87C3G, a Novel, Truncated C3G Isoform that is Overexpressed in Chronic Myeloid Leukemia and Interacts with BCR–ABL

Gutierrez-Berzal J, Castellano E, Martin-Encabo S, Gutierrez-Cianca N, Hernandez JM, Santos E, Guerrero C

Centro de Investigacion del Cancer, IBMCC, Universidad de Salamanca-CSIC, Campus Miguel de Unamuno, 37007 Salamanca, Spain

C3G is a Crk binding protein showing transformation suppressor activity towards several oncogenes through a GEF-independent mechanism. Here, we describe a novel C3G isoform, designated p87C3G, lacking the most amino terminal region of the cognate protein that is to be overexpressed in two CML cell lines, K562 and Boff 210, both expressing Bcr–Abl p210. p87C3G expression is also highly augmented in patients diagnosed with chronic myeloid leukemia (CML) Ph+, in comparison with healthy individuals, and returns to basal levels after treatment with STI571. p87C3G coimmunoprecipitates with both CrkL and Bcr–Abl in CML cell lines and coimmunoprecipitation between p87C3G and Bcr–Abl was also detected in primary cells from CML patients. These interactions have been confirmed by in vitro pull down experiments. The interaction between p87C3G and Bcr–Abl involves the SH3-binding domain of p87C3G and the SH3 domain of Abl and depends mostly on the first polyproline region of p87C3G. Furthermore, p87C3G is phosphorylated in vitro by a Bcr–Abl-dependent mechanism, probably in complexes with Hck. These results indicate that p87C3G overexpression is linked to CML phenotype and that p87C3G may exert productive functional interactions with Bcr–Abl signaling components suggesting the implication of this C3G isoform in the pathogenesis of chronic myeloid leukemia.

Characterization of C3G Suppression Mechanism

Martín-Encabo, S. Santos, E; Guerrero, C

Centro de Investigacion del Cancer, IBMCC, Universidad de Salamanca-CSIC, Campus Miguel de Unamuno, 37007 Salamanca, Spain

The guanine nucleotide exchange factor (GEF) C3G is an activator of *Rap1*, a nononcogenic member of the Ras family. We have demonstrated that C3G suppresses oncogenic transformation induced by several oncogenes (*Ras, c-Sis, v-raf, dbl y R-Ras*) independently of its catalytic domain responsible of *Rap1* activation. In Hraslys12-transformed NIH3T3 overexpression of C3G or C3GDCat (mutant lacking the catalytic domain) inhibited Hraslys12-mediated phosphorylation of ERK without affecting the rest of the components of the Ras-ERK cascade. We have evidences that the fraction of C3G involved in transformation suppression is restricted to the subcortical actin cytoskeleton as (i) mutant C3GDCat localizes to this area and is not presented in the Golgi apparatus, (ii) C3G directly interact with actin through the SH3-domain (iii) the cytoskeleton disruption with cytochalasin D reverts the C3G inhibitory effect on ERK phosphorylation. Also our results show that Ser/Thr phosphatase (PP2A), could be a possible mediator of the C3G suppressor activity. Overexpressed C3G and C3GDCat cells have more phosphatase activity global and associated with phosphorylated ERK, the interaction between phosphorylated MEK and PP2A is constitutive, PP2A is found in C3G inmunoprecipitates. Our hypothesis is that C3G may stimulate the binding of PP2A to MEK in the complexes where ERK is located, favoring its dephosphorylation.

Differential Gene Expression Profile of Pancreatic Islets in GRF1-Deficient Mice

M. Arribas, A. Nunez, J. Rivas, and E. Santos

Instituto de Biologia Molecular y Celular del Cancer, IBMCC (CSIC-USAL). University of Salamanca, 37007 Salamanca, Spain

The *Grf* gene family codes for guanine nucleotide exchange factors activating Ras proteins. *Grf2* knockout mice do not show any phenotype. *Grf1*-deficient mice exhibit reduced body weight, hypoinsulinemia and reduction of beta-cell mass. We compared the genomic expression profiles of wild-type and Grf1KO pancreatic islets. Wild type and Grf1KO RNA samples were hybridized to Affymetrix oligonucleotide microarrays. Pair-wise or global statistical comparisons allowed identification of 29 differentially expressed genes. qRT-PCR, western blot and inmunohistochemistry assays have been used to confirm the RNA expression changes previously identified. Functional classification of the genes pointed to potential roles of Grf1 in calcium signaling in pancreatic beta cells. 14 genes were increased in *Grf1* knockout islets, among them Slco1a6 (solute carrier organic anion transporter 1a6) and Gria (glutamate receptor ionotropic 2) a regulator of intracellular calcium. The list of repressed genes identified, included Iapp (Islet amyloid polypeptide), a locus implicated in regulation of insulin secretion, and enzymes such as Gdpd3 (glycerophosphodiester phosphodiesterase domain containing 3) and Pah (phenylalanine hydroxylase). Our data document the occurrence of specific transcriptomic changes in pancreatic islets of *Grf1* knockout mice and point to a unique functional role of *Grf1* as modulator of Ras activation in islet cells.

MET Activation and Fumarase Suppression Act Sinergistically in Conferring a Motile Phenotype to Mouse Embryo Fibroblasts

Barbara Costa[1], Luigi Ferrantino[1], Marzia Armaro[1] and Maria Flavia Di Renzo[1]

[1] *Laboratory of Cancer Genetics of the Institute for Cancer Research and Treatment (IRCC), University of Torino School of Medicine, 10060 Candiolo, Italy*

The role of MET in human tumor formation has been established after the discovery of germline activating mutations in patients suffering from hereditary papillary renal cell carcinomas. On the other hand MET activation by overexpression, is frequently found in primary human cancers of specific histotype, including sporadic kidney carcinoma. Although MET activation by either mutation or overexpression plays a crucial role in the onset of kidney carcinoma other genetic or epigenetic events must contribute.

Papillary renal cell carcinomas are also associated to germline inactivating mutations of the gene encoding the mitochondrial enzyme fumarate hydratase (FH). The mechanism through which loss of FH function leads to tumor formation remains to be determined although several findings suggest that mitochondria participation in apoptosis and proliferation control is critical. Aim of this work is to study the association between MET activation and FH deficiency in cell transformation and tumor progression.

To this aim, we used lentiviral vector to stably express, in primary cultured mouse embryo fibroblasts, both an activated form of MET oncogene and a short hairpin RNA specific for the FH. Results show that the concomitant MET expression together with the suppression of FH activity act synergistically in conferring motility to the cells, a property that is peculiar of transformed cells.

Methylation of GSTP1 Promoter in Prostate Cancer as a Target for Combination Treatment of Hypmethylating Agents and Brosallicin

Maria Antonietta Sabatino, Massimo Broggini

Laboratory of Molecular Pharmacology, Departement of Oncology, Istituto Mario Negri, Milan, Italy

Human prostatic cancer is characterized by the methylation of GSTP1 promoter, explaining for the absence of the enzyme glutathione-*S*-transferase (GST) pi; this characteristic is retained by the prostatic cell line, LNCaP. LNCaP cells have been treated with hypomethylating drugs, in order to re-express the GST-pi enzyme, important for the cytotoxicity of Brostallicin, whose activity depends on intracellular levels of glutathione/glutathione-*S*-transferase-pi. Brostallicin alone, was inactive against LNCaP cells, (MTT assay: IC50 = 250 ng ml^{-1}), while is active on LNCaP cells transfected with GST-pi cDNA (IC50 = 50 ng ml^{-1}). 5-aza-deoxycytidine resulted toxic on LNCaP cells and the activity of the combination with Brostallicin could not be appreciated. The less toxic cytidine analog Zebularine was used (100—125 µM) to pretreat LNCaP cells (96—120 h) before treatment with Brostallicin; the cytotoxicity of the combination resulted significantly higher than the single agents. GST enzymatic activity was detectable in cells treated with Zebularine, but not in the controls. Moreover, Zebularine was also well tolerated in vivo (on SCID mice), in prolonged treatment. These results represent the basis for further studies of this combination in vivo, and they could have important clinical relevance for treatment of prostate cancer.

Biological Activities of Chemokines and their Receptors in Thyroid Cancer

Rosa Marina Melillo[1], Valentina Guarino[2], Valentina De Falco[2], Fulvio Basolo[3], Pinuccia Faviana[3], Riccardo Giannini[3], Giuliana Salvatore[2], Tito Nappi[2], Paolo Salerno[2], Massimo Santoro and Rosa Marina Melillo[2]

[1] *IEOS, DBPCM, Napoli Italy*
[2] *Istituto di Endocrinologia ed Oncologia Sperimentale del CNR G. Salvatore/Dipartimento di Biologia e Patologia Cellulare e Molecolare, University Federico II, 80131 Naples, Italy*
[3] *Dipartimento di Oncologia, Pisa, Italy*

Epithelial thyroid tumors include well-differentiated papillary and follicular (PTC and FTC), poorly differentiated (PDC), and anaplastic carcinomas (ATC). We showed that transformation of thyroid cells with oncogenes typical of PTCs activates a transcriptional program that includes the expression of chemokines, and of their receptors. Chemokines were described originally to regulate the chemiotaxis of leukocytes into inflammatory sites. Furthermore, chemokines can be secreted by tumor cells and recruit, into the tumor site, inflammatory cells. To exploit the eventual role of these molecules in thyroid cancer, we screened human thyroid cancer-derived cells and human thyroid tumor samples for the expression of the chemokine receptors CXCR2 and CXCR4, and their ligands, CXCL1, IL-8, and SDF-1. We found that both well differentiated (PTC) and undifferentiated (ATC) thyroid cancers expressed the chemokine receptors CXCR2 and CXCR4. Most of the PTC and some ATC-derived cell lines also secreted the chemokines CXCL1 and IL-8, but not SDF-1. We show here that chemokines treatment of human thyroid cancer-derived cells can mediate cell proliferation, apoptosis and chemoinvasive ability. Based upon these results, we are currently testing the possible use of small compounds with inhibitory activity toward CXCR2 and CXCR4 as novel anticancer therapies for ATC.

Mass Spectrometry Identification of Cyclin B1 Serine 133 as Unique and Specific Phospho-Site for Polo-Like Kinase-1

Simona Rizzi, Barbara Valsasina, Sonia Troiani, Rosario Baldi, Henryk Mariusz Kalisz, Luisa Rusconi, Simon Plyte and Antonella Isacchi

Nerviano Medical Sciences, Biotechnology, Nerviano, Milano Italy
Discovery Research Oncology, Nerviano Medical Sciences, 20014 Nerviano (MI) Italy

The cyclin-dependent kinase 1 (Cdk1)/cyclin B1 complex plays crucial roles in eukaryotic mitotic progression. During S phase and G2 the complex is mostly localized in the cytoplasm, but at the G2/M transition it translocates to the nucleus and this nuclear accumulation is promoted by the phosphorylation of four serine residues in Cyclin B1. Polo-like kinase 1 (Plk1) that plays different roles in the cell cycle and has been implicated in several types of cancer, colocalizes with Cdk1/Cyclin B1 complex during late prophase and mitosis. It is also reported that Cyclin B1 is a good substrate for Plk1. To investigate which are the major Plk1 phosphorylation site(s) on Cyclin B1, we in vitro phosphorylated recombinant Cyclin B1 with Plk1. Cyclin B1 resulted stochiometrically monophosphorylated by Plk1 and MS/MS analysis showed that Serine 133 was the unique site of phosphorylation. In addition, phosphorylation of Cyclin B1 with Plk2 and Plk3, two other members of the Polo-like kinase family, showed only minor levels of phosphorylation at Serine 133. These results suggest that phosphorylation of Cyclin B1 at Serine 133 is specific for Plk1, among the Plk kinases, and could be a starting point to develop a specific Plk1 marker to follow Plk1 activity in cells.

Suppression of p53 by Notch in Lymphomagenesis: Implications for Initiation and Regression

Levi J. Beverly*, and Anthony J. Capobianco*

*The Wistar Institute, Molecular & Cellular Oncogenesis Program, Philadelphia, PA 19104, USA

Aberrant Notch signaling contributes to more than half of all human T-cell leukemias, and accumulating evidence indicates Notch involvement in other human neoplasms. We developed a tetracycline inducible mouse model (Top-Notchic) to examine the genetic interactions underlying the development of Notch-induced neoplastic disease. Using this model we demonstrate that Notch suppresses p53 in lymphomagenesis through repression of the ARF–mdm2–p53 tumor surveillance network. Attenuation of Notch expression resulted in a dramatic increase in p53 levels that led to tumor regression by an apoptotic program However, all tumors relapsed with rapid kinetics. Furthermore, by directly inhibiting the mdm2–p53 interaction by using either ionizing radiation or the novel small molecule therapeutic Nutlin, p53 can be activated and cause tumor cell death, even in the presence of sustained Notch activity. Therefore, it is the suppression of p53 that provides the Achilles heel for Notch induced tumors, as activation of p53 in the presence of Notch signaling drives tumor regression. Our study provides proof of principle for the rational targeting of therapeutics against the mdm2–p53 pathway in Notch-induced neoplasms. Furthermore, we propose that suppression of p53 by Notch is a key mechanism underlying the initiation of T-cell lymphoma.

Author Index

A

Abraham, R.T., 48–49
Abrahamsson, P.A., 97
Acquadro, F., 203, 206
Adams, M.M., 51, 55
Adema, G.J., 76
Adini, A., 72
Agami, R., 17, 18–19, 21, 26, 28, 40, 42
Aglietta, M., 201
Aguilera, A., 198
Aharinejad, S., 72, 78
Aitken, A., 158–159
Alblas, J., 98
Alcalay, M., 176, 230
Alessandrini, I., 231
Alessandro, F., 191
Alessandro, R., 191
Allavena, P., 68, 72, 74, 79, 195
Alloisio, M., 217
Almstrup, K., 29
Amati, B., 118, 184, 193, 202
Ambosio, M., 221
Amiard, S., 218
Ammendola, V., 221
Anastasi, A.M., 183
Arcuri, M., 221
Arden, S.R., 138
Armaro, M., 249
Armellin, M., 238
Arribas, M., 248
Aster, J.C., 130
Aurisicchio, L., 204
Avruch, J., 159
Azenshtein, E., 71

B

Babarovic, P., 245
Baert, J.-L., 177
Bagga, S., 18
Bailis, J.M., 6
Bakkenist, C.J., 47–48
Baldi, R., 252
Balkwill, F., 57, 64, 67–68, 70, 79
Balladore, E., 224
Ballarini, M., 196
Bambara, R.A., 4
Bamberger, C., 240
Barabasz, A., 138, 139
Bardelli, A., 201
Bar-Eli, M., 88
Baryshnikova, E., 224, 234
Basolo, F., 251
Bauwens, S., 218
Beaudoin, C., 177
Bell, D., 72
Bellini, A., 221
Bennett, S., 118, 184
Berchuck, A., 74
Bergsma, D.J., 3, 9
Bernard, L., 193, 230
Bernards, R., 42
Bettolini, R., 223
Beverly, L.J., 253
Biamonti, G., 232
Bianchi, P., 217
Bingle, L., 71–72
Biswas, S.K., 69, 74
Blandino, G., 215, 226, 237
Blasco, M.A., 115

Blow, J.J., 5–7
Bochar, D.A., 50
Bongiorno, G., 182
Borowiec, J.A., 4
Borrello, M.G., 78
Bosari, S., 217
Bossi, G., 226
Bossi, M., 205
Bottazzi, B., 69
Bours, V., 188
Braig, M., 18, 34
Brancaccio, T., 221
Bratt, A., 192
Brehm, A., 229
Breiling, A., 244
Brenner, S., 9
Bridges, D., 159
Broggini, M., 199, 250
Bronte, V., 76–77
Brummelkamp, T.R., 40
Bruno, T., 186
Brunori, M., 218
Bucci, G., 196
Burger, H., 33
Burgers, P.M., 4
Burkhart, R., 5
Bylund, G.O., 4

C

Cabez, T., 151, 167
Calautti, E., 214
Califice, S., 74
Cambiaghi, V., 180
Canevari, S., 197
Caparros, E., 200
Capobianco, A.J., 253
Capra, M., 63, 243
Carbon, J., 9
Carbone, R., 182
Carrassa, L., 199
Carta, L., 78
Castellano, E., 246
Castellano, G., 197
Castermans, E., 188
Castronovo, V., 74
Caux, C., 75
Cavalloni, G., 201
Celis, J.E., 151, 167
Cera, M.R., 205

Ceresoli, G.L., 217
Cernilogar, F.M., 244
Cesaroni, M., 178
Chai, Y.L., 47
Chan, L.C., 108
Chariot, A., 188
Chatterton, R.T. Jr., 185
Chen, C.Z., 19
Chen, Z., 18
Chiarle, R., 131
Chimenti, S., 205
Chiocca, S., 230, 245
Chiodoni, C., 231
Chiorino, G., 203, 206, 214
Chong, J.P., 5
Christofori, G., 104
Ciavolella, A., 137
Ciliberto, G., 204
Cipriani, B., 204
Cirillo, A., 204
Ciro, M., 63
Clark, S.S., 108
Clemons, P.A., 138
Clerici, E., 182
Close, P., 188
Cohen, G.M., 142
Collado, M., 18
Collins, K.L., 4
Colloca, S., 204, 221
Colombo, E., 227
Colombo, M.P., 79, 231
Comoglio, P.M., 179
Conti, I., 70, 71
Corada, M., 205
Corbi, N., 186
Cordani, N., 217
Cornelius, L.A., 71
Corso, S., 179
Cortese, R., 189, 204, 221
Costa, B., 249
Costanzo, V., 7
Coussens, L.M., 67–68, 75
Crescenzi, M., 186
Croce, C.M., 116
Crum, C.P., 240
Cuadrado, M., 62
Cundari, E., 189
Curiel, T.J., 76
Cvetic, C., 5

D

D'Adda di Fagagna, F., 190
D'Alessandris, G., 209
Dall' Olio, V., 193
Damia, G., 199
Danesi, R., 242
D'angelo, C., 186
Danis, E., 6–7, 9
Dassa, E., 225
Dave, S., 130
David-Pfeuty, T., 243
De Angelis, F.G., 191
De Angelis, N., 205
De Antoni, A., 184
De Bortoli, M., 87
De Braud, F., 242
Decker, S.J., 98
De Falco, V., 251
De Filippo, K., 87
Degerny, C., 177
De Herder, W.W., 98
Deininger, M., 108
Dejana, E., 205, 211
De Launoit, Y., 177
De Leval, L., 188
Dellino, G., 149
Delia, D., 186
Del Sal, G., 210
Demontis, S., 238
De Nicola, F., 186
De Pra, A., 221
Derbyshire, D.J., 51
De Rinaldis, E., 189
De Rinaldis, M., 221
De Rosa, G., 179
De Santis, G., 197
De Santis, R., 183
Desnues, B., 69
Destro, A., 217, 224, 234
Devault, A., 5
De Visser, K.E., 79
De Vita, G., 17, 26, 40
DeVries, T.A., 157
Di Agostino, S., 215, 237
Dierov, J., 162
Diffley, J.F., 5
DiFiore, P.P., 173
Dimiziani, L., 186
Dinapoli, M.R., 74

Dinarello, C.A., 69
Ding, G.J.F., 138
Di Padova, M., 186
Di Renzo, M.F., 225, 249
Di Vignano, A.T., 214
Doench, J.G., 18
Dominguez, M., 200
Dong, Z., 75
Donovan, S., 5
Dornreiter, I., 4
Dotto, G.P., 214
Dozio, E., 96, 194
Draetta, G.F., 230
Droege, M., 151, 168
Drost, J., 17
Du, L.L., 51
Durant, S.T., 50
Durocher, D.H.J.F.A.R., 49
Dutta, A., 5–6
Duursma, A., 40
Duyndam, M.C., 72
Dyson, N., 229

E

Ebert, B.L., 130
Edwards, M.C., 5–6
El-Deiry, W.S., 29
Elledge, S.J., 49
Elvin, J.A., 240
Emili, A., 49
Emiliozzi, V., 215
Ercole, B.B., 204
Evan, G.I., 228

F

Fagiani, E., 243
Falchetti, M.L., 209
Falleni, M., 217
Fanciulli, M., 186
Fassetta, M., 87
Faviana, P., 251
Felici, A., 211
Feng, W., 87
Fernandez, P.L., 74
Fernandez-Capetillo, O., 62
Ferrantino, L., 249
Filocamo, G., 189
Finch, A., 228
Fingar, D.C., 142

Finn, O.J., 174
Finocchiaro, G., 193
Fiordaliso, F., 205
Fiorenzo, P., 209
Floridi, A., 186
Foglietti, C., 189
Foray, N., 51
Forsburg, S.L., 5–6
Forstemann, K., 39, 40
Franchi, G., 224
Franchimont, N., 188
Friis, E., 151
Fukasawa, K., 34
Fumagalli, M., 190
Funa, K., 181, 235

G

Gabrilovich, D.I., 74
Gai, D., 4
Gallego, C., 98
Gallinari, P., 207
Galvani, A., 137
Ganzinelli, M., 199
Gardini, A., 176
Garg, P., 4
Gargiulo, G., 196
Garlanda, C., 80
Garnier, F., 74
Garziera, M., 238
Gasparri, F., 137–138, 140–142
Gennari, L., 224
Gerald, C., 97
Germano, G., 243
Ghassabech, G.H., 68
Giampieri, S., 241
Giannini, R., 251
Giardina, G., 243
Gibbs, J.B., 140
Gilbert, C.S., 52
Gilbert, D.M., 9
Gillespie, P.J., 5
Gillis, Ad J.M., 17
Gilson, E., 218
Giordano, S., 179
Giorgetti, L., 182
Giorgio, M., 178
Giraudo, E., 79
Giraud-Panis, M.J., 218
Giuliano, K.A., 137, 138

Goding, C., 59
Goerdt, S., 72
Goldberg, Y., 97
Golub, T.R., 130
Gonchoroff, N.J., 142
Gordon, E.A., 97
Gordon, S., 68
Gorrini, C., 118
Grazini, U., 63, 236
Gregory, P.D., 196
Griekspoor, A., 17
Grierson, D.S., 243
Grizzo, A., 238
Groffen, J., 108
Gromov, P., 151, 167
Gromova, I., 151, 167
Grossi, M., 214
Gu, L., 76
Guana, F., 203
Guarino, V., 251
Guccione, E., 193
Guerrero, C., 246, 247
Guiducci, C., 77
Gutierrez-Berzal, J., 246
Gutierrez-Cianca, N., 246
Gutierrez-Garcia, I., 200

H

Hagemann, T., 75
Haghnegahdar, H., 71
Hahn, W.C., 18, 26–27, 150, 166
Haley, B., 18
Hanahan, D., 26, 38
Hanamoto, H., 71
Hannon, G.J., 238
Hansel, D.E., 97
Hansen, E.F., 167
Hantschel, O., 156
Hantusch, B., 104
Harland, R.M., 9
Hartwell, L.H., 5, 47
Harvey, K.J., 7
Hatfield, S.D., 40
Hay, R.T., 122
He, L., 18
Helin, K., 63, 223
Hendzel, M.J., 142
Hernandez, J.M., 246
Herzog, H., 97

Hilger-Eversheim, K., 87–88
Hiou-Feige, A., 214
Holmgren, L., 192
Hotchkiss, A., 75
Hsiao, C.L., 9
Hu, Y., 109
Huang, S., 88
Huang, Y., 161
Huberman, J.A., 5
Hubscher, U., 4
Hughes, M., 222
Hui, L., 212
Hungerford, D., 108
Hurwitz, J., 4, 6
Huyen, Y., 47, 52
Hyrien, O., 6, 9

I

Iezzi, S., 186
Iizuka, M., 7
Ilaria, R.L., 112
Imhof, A., 244
Infante, M., 234
Inghirami, G., 129
Irene, B., 191
Isaacs, J.T., 78
Isacchi, A., 252
Ishimi, Y., 7
Iyer, L.M., 6

J

Jackson, A.L., 130
Jackson, S.P., 47, 49
Jacob, F., 9
Jaffe, E.S., 71
Jares, P., 5, 7
Jarnicki, A.G., 76
Jekimovs, C., 186
John, B., 18
Johnson, S.M., 18
Jones, P.A., 121
Joseph, I.B., 78
Josèphe, M., 218
Justice, R.W., 34

K

Kalisz, H.M., 252
Kaneko, M., 181
Kantarjian, H., 108

Kao, H.I., 4
Kaplan, D.L., 7
Karin, M., 124
Kastan, M.B., 47, 48
Keffel, S., 97
Kelly, T.J., 4
Kerjaschki, D., 104
Kersemaekers, A.M., 29, 33
Khanna, K.K., 186
Kharbanda, S., 156
Kim, D.S., 98
Kim, H.Y., 4
Kim, S.T., 48
Kirschner, M.W., 5
Kleinhans, M., 71
Klimp, A.H., 74
Koberle, B., 29
Kolfschoten, I.G., 18, 27, 41–42, 130
Koonin, E.V., 6
Korenjak, M., 229
Korner, U., 7
Kortylewski, M., 73
Kouno, J., 71
Kozlov, S., 48
Kubota, Y., 5, 7
Kufe, D., 157
Kusmartsev, S., 74
Kutok, J.L., 130

L

Labib, K., 5, 7
Laghi, L., 224
Lahm, A., 189, 207, 221
La Monica, N., 204
Lanfrancone, L., 243
Lang, P., 137, 138
Lapi, E., 226
Larghi, P., 67
Larhammar, D., 97
Larocca, L.M., 209
Laskey, R.A., 6, 9
Latini, R., 205
Leask, A., 88
Lecis, D., 186
Lee, J.K., 6, 7
Lee, S., 214
Lee, S.E., 47
Legraverend, M., 243

Lehembre, F., 104
Lei, M., 5
Lenain, C., 218
Leone, B.E., 195
Leone, F., 201
Leonetti, C., 186
Leonhardt, H., 4
Levi, A., 209
Levine, A.J., 39
Levy, S., 196
Lewis, B.P., 18–19, 42
Li, A., 71
Li, H., 50
Li, Y., 34
Lim, L.P., 18, 34
Lin, E.Y., 72
Liu, D., 158
Liu, Y., 50
Liu, Y.-P., 17
Livingston, D.M., 175
Locati, M., 68, 75
Looijenga, L.H., 29, 39
Looijenga, L.H.J., 17
Lorenzato, A., 225
Lou, Z., 49, 50
Lough, J., 118
Lövborg, H., 138
Lowe, S.W., 107
Lowndes, N.F., 47
Lu, J., 39
Ludwig, O., 243
Lundgren, M., 3
Lutzker, S.G., 39
Luzi, L., 178, 193, 243
Luzzago, A., 221

M
Mabuchi, S., 98
Mack, D., 97–98
Madine, M.A., 6
Maestro, R., 238
Magni, P., 96–97, 194
Mahbubani, H.M., 5–6
Mahon, C., 3
Maiorano, D., 5
Maira, G., 209
Malaise, M., 188
Malesci, A., 224
Mancino, A., 67

Mantovani, A., 67–72, 74–77, 79, 103, 195
Mantovani, F., 210
Marangi, I., 182
Marchesi, F., 195, 233
Maric, C., 9
Martinato, F., 193
Martinelli, P., 227
Martin-Encabo, S., 246–247
Martinez, B., 62
Mascheroni, D., 178
Masters, J.R., 29, 39
Matias, S., 239
Matsumoto, T., 4
Matsumoto, Y., 181
Matsushima, K., 69
Mauen, S., 177
Mayer, F., 29
Mazzi, M., 197
McBlane, F., 63, 236
McGarry, T.J., 5
Mckeon, F., 240
McPherson, J.P., 34
McPherson, L.A., 88
Melillo, R.M., 251
Mello Grand, M., 203, 206
Meola, A., 204
Merville, M.-P., 188
Michaelis, A.H., 240
Michaloglou, C., 18
Michel, M.C., 97
Migliaccio, E., 178
Migliore, C., 179
Milani, P., 182
Mills, A.D., 6
Minardi, S., 230
Minenkova, O., 183
Minguez, J.M., 138
Minucci, S., 196, 239
Miranda, E., 224
Mochan, T.A., 49
Monaci, P., 221
Monestiroli, S., 233
Mongiardi, M.P., 209
Monica, B., 191
Montagnoli, A., 7
Montano, N., 209
Monte, D., 177
Montecucco, A., 232

Monteriu, G., 183
Monti, O., 237
Monti, P., 71, 195
Moorhead, G.B., 159
Moreira, J.M.A., 151, 167
Morenghi, E., 217, 224
Mori, K., 71
Moroni, M.C., 223
Moser, M., 88
Mostert, M., 39
Motoike, T., 205
Motta, M., 96–97, 194
Mottolese, M., 215
Muller, A.J., 79
Mullins, D.E., 97
Murakami, Y., 4
Murga, M., 62
Musacchio, A., 184, 220
Muslin, A.J., 159

N
Naeem, V., 185
Nagakawa, O., 98
Nagar, B., 156
Nagel, R., 17
Nakamura, H., 7
Nappi, T., 251
Nardi, C., 207
Nasheuer, H.-P., 3
Nasi, S., 228
Nesbit, M., 71
Neuwald, A., 6
Newport, J., 5, 7
Nguyen, V.Q., 5
Nickoloff, J.A., 50
Nicosia, A., 204, 221
Nie, M., 97
Noel, W., 69
Nojima, H., 17
Nomura, M., 160
Nowell, P., 108
Nowicki, A., 72
Nunez, A., 248
Nyormoi, O., 88

O
O'Donnell, K.A., 18
Oehlmann, M., 3, 5–6
Oku, T., 4
Olga, S., 191
Olivero, M., 225
Olivier, S., 188
O'Neill, T., 48
Oosterhuis, J.W., 29, 39
Orfanos, C.E., 72
Orlando, V., 244
Orsatti, L., 221
Orso, F., 87
Ostano, P., 214
Ota, A., 18
Ouchi, T., 49

P
Pacchiana, G., 63
Paik, S., 67, 68
Pallaoro, M., 207
Pallini, R., 209
Palombo, F., 221
Paolini, C., 207
Parazzoli, D., 205
Parenza, M., 231
Parsons, R., 4
Passananti, C., 186
Pavlov, Y.I., 4
Pavoni, E., 183
Payne, A.S., 71
Pear, W.S., 109
Pedrazzini, T., 96–97
Peeper, D.S., 219
Pelicci, P.G., 178, 180, 182, 196, 227, 233
Pellegrini, C., 217
Pellegrino, E., 129
Pellieux, C., 97
Penachioni, J.Y., 201
Pendergast, A.M., 156
Peng, M., 50
Penna, E., 87
Perlman, Z.E., 139, 146
Perna, D.F., 202
Perro, M., 225
Peruzzi, D., 204
Pestryakov, P.E., 23
Petronzelli, F., 183
Pevarello, P., 141
Phillips, R.J., 78
Piaggio, G., 215
Piatti, S., 220

Piccinin, S., 238
Piemonti, L., 195
Pignochino, Y., 201
Pillai, R.S., 18
Piseri, P., 182
Piva, R., 129, 132
Plyte, S., 252
Pollard, J.W., 67, 72, 78
Polo, S., 208
Porrello, A., 186
Porta, C., 67
Prado, F., 198
Prieur, A., 219
Prokhorova, T.A., 7
Prosperini, E., 63
Pui, C.-H., 109
Pulford, K., 130, 131
Pumiglia, K.M., 98

Q
Quarto, M., 63, 243

R
Rada, E., 233
Radisky, D., 60
Ramm, P., 138
Randle, D.H., 109
Rank, F., 151, 167
Rappold, I., 49
Rasiah, K.K., 97–98
Rauh, M.J., 77
Ren, R., 109
Reubi, J.C., 98
Reuther, G.W., 162
Riabowol, K., 222
Richards, G.R., 138
Richmond, A., 70
Riggs, A.D., 5
Rinaldo, C., 226
Risio, M., 201
Rittinger, K., 163
Rivas, J., 248
Rizzi, A., 217
Rizzi, S., 252
Robins, H., 18
Roche, K.C., 47
Rollins, B., 70
Rollins, B.J., 70
Romanenghi, M., 196

Roncalli, M., 217, 224, 234
Rosenquist, M., 158
Rossi, R., 232
Rouse, J., 47
Rowles, A., 5–6
Rowley, J.D., 108
Rubino, L., 67
Ruckes, T., 71
Rudini, N., 211
Ruscica, M., 96–98, 194
Rusconi, L., 252
Russell, M., 222
Rustighi, A., 210
Rustin, P., 225

S
Sabatini, D.M., 61
Sabatino, M.A., 250
Sacchi, A., 215, 226, 237
Sage, C.L., 17
Sahai, E., 241
Saji, H., 71
Sakaguchi, S., 71, 76
Salerno, P., 251
Salio, M., 205
Salmon, E.D., 220
Salvatore, G., 251
Sanders, S.L., 52
Sangaletti, S., 79
Sanges, R., 202
Santapaola, D., 183
Santini, C., 221
Santoro, M., 251
Santos, E., 246
Santos, S., 247
Sardella, D., 118
Sarotto, I., 201
Satterlee, J., 229
Savino, M., 228
Scala, S., 71
Scanziani, E., 233
Scarpino, S., 72
Scarsella, M., 186
Scatolini, M., 203, 206
Scheuch, H., 212
Schoppmann, S.F., 74
Schorle, H., 88
Schriemer, D., 222
Schrier, M., 17

Schutyser, E., 71, 76
Schwartz, M.F., 49
Scita, G., 187
Scotton, C.J., 69
Sehnke, P.C., 158
Seiser, C., 230
Selbie, L.A., 97
Senese, S., 230
Sengupta, S., 47
Serrano, M., 18, 22
Sessa, C., 78
Shaffer, A.L., 130
Shah, N.P., 108
Shakes, D.C., 158
Sharp, P.A., 18
Sherr, C.J., 107, 141
Shiekhattar, R., 123
Shing, D.C., 233
Sica, A., 67–68, 74–75, 77
Silva, T., 7
Simmons, D.T., 4
Sinha, I., 192
Sinha, P., 77
Sismondi, P., 87
Smellie, A., 138
Smith, A.P., 202
Smith, J.R., 71
Smith, R., 130
Snyder, M., 6
Soddu, S., 186
Solero, A., 87
Sonego, M., 238
Soucek, L., 228
Soza, S., 232
Sozzani, S., 68
Squatrito, M., 118
Staszewsky, L.I., 205
Staudt, L.M., 130
Steinkuhler, C., 189, 207
Stendardo, M., 233
Stinchcomb, D.T., 9
Stoop, H., 17
Strano, S., 215, 226, 237
Stucki, M., 50
Stupka, E., 202
Subhra Kumar, B., 67
Suh, E.-K., 240
Suh, M.R., 29, 33

Sulli, G., 223
Sunayama, J., 159, 160
Suyama, E., 88
Suzuki, K., 48

T
Taagepera, S., 156
Tabruyn, S., 188
Tacchetti, C., 169, 213
Tada, S., 5
Tainio, H., 97–98
Takahashi, H., 69
Takahashi, T.S., 7
Takahashi, Y., 34
Talamo, F., 221
Talpaz, M., 108
Tanaka, S., 5
Tania, I., 191
Tarapore, P., 34
Taverna, D., 87
Taylor-Harding, B., 229
Teichmann, M., 71
Teruya-Feldstein, J., 71
Testa, G., 216
Tibbetts, R.S., 48
Tizzoni, L., 193
Tjian, R., 87
Tlsty, T.D., 101
Tocco, F., 210
Toffalorio, F., 242
Toji, S., 34, 41
Tomassetti, A., 197
Torroella-Kouri, M., 77
Tosato, G., 71
Trent, J.M., 142
Troiani, S., 252
Trowbridge, P.W., 4
Trubia, M., 233
Trumpp, A., 202
Tsuda, Y., 69
Tsuruta, F., 159, 160
Tuteja, N., 6
Tuteja, R., 6
Tzivion, G., 159

U
Ueno, T., 70, 71
Urnov, F., 196
Uyttenhove, C., 79

V

Vago, R., 232
Vaira, V., 217
Valsasina, B., 252
Van Damme, J., 71
Van den Brule, F., 74
Van Duijse, J., 17
Van Hatten, R.A., 5
Vecchi, A., 68, 205
Venesio, T., 203
Ventura, S., 97
Verma, I.M., 47, 58
Vermeer, M.H., 71
Vialard, J.E., 49
Vicari, A.P., 75
Vigneri, P., 162
Villa, A., 205
Vinci, M., 205
Viniegra, J.G., 117
Vitelli, A., 221
Voitenleitner, C., 3
Voorhoeve, P.M., 17, 18–19, 21, 26, 28, 40, 42

W

Waga, S., 4–6
Wagner, E.F., 212
Walter, J., 5
Walter, J.C., 5, 7
Wan, W., 133
Wang, J.Y., 156, 158, 162
Wang, T., 73
Wang, X., 88
Warbrick, E., 4
Ward, I.M., 47
Watt, F., 181
Weinberg, R.A., 26, 28
Weinert, T., 29
Weisshart, K., 4
Wellinger, R.E., 198
Werb, Z., 67
Werling, U., 88
Westbrook, T.F., 18, 130
Westermark, U.K., 51
Wetterskog, D., 235
Whitfield, J., 228
Wick, N., 104

Wicki, A., 104
Williams, R.T., 107, 110, 112
Williams, T., 87
Witte, O.N., 108–109
Wolff, N.C., 112
Won, J., 160
Wong, S., 109
Woodward, A.M., 6
Wright, J.A., 49
Wynn, T.A., 69
Wysocki, R., 52

X

Xiao, B., 158
Xing, H., 159
Xu, T., 34

Y

Yabuta, N., 17
Yang, A., 240
Yang, J., 70
Yang, W., 181, 235
Yanow, S.K., 7
Yi, F., 71
Yoshida, K., 155, 157, 162
Yoshimura, T., 69
Yoshizawa-Sugata, N., 7
You, Z., 7
Yuan, Z.M., 162

Z

Zamore, P.D., 18
Zanardi, A., 182
Zannini, L., 186
Zanon, C., 201
Zaragoza-Dorr, K., 230
Zembutsu, A., 5, 6
Zerbini, V., 215
Zhang, J., 51
Zhang, X., 109
Zhao, F., 87
Zhu, J., 156
Zhu, Z., 69
Zlotorynski, E., 17
Zou, L., 49
Zucali, P.A., 217

Subject Index

A

Abl-mediated apoptosis, 160
Abl tyrosine kinase
  dissociation of, 159
  in DNA damage response, 157
  inhibitors, 108
  nuclear targeting of, 157, 160
  nuclear translocation of, 162
  oncogenic forms of, 156
  phosphorylation of, 159
  structure and function of, 156
Acetyl-transferase Tip60, in Myc-induced DDR, 118
Actin binding proteins, 187
Actin cytoskeleton, 59
Actin dynamics-based cell migrations, 187
Activator protein 1, in liver cancer development, 212
Acute lymphoblastic leukemias, 107
Acute myeloid leukemia (AML), 176, 180, 233
Acute promyelocytic leukemia, 180
Adenovirus (Ad), 204
Adherens junctions (AJ), 211
Amplicons, ultra deep sequencing of, 152, 168
Anaplastic large cell lymphomas (ALCL), 129, 130
  ALK expression signature in, 132–134
  neoplastic phenotype, 134
  pathogenesis of, 131
Anaplastic lymphoma kinase (ALK), 129, 130
  catalytic domain of, 131
  knockdown cell lines, 131–132
  signature, functional validation of, 134
Angiogenesis, 58, 74
  control of, 192
  induction of, 211
Angiomotin, in endothelial cell–cell contacts, 192
Angiostatin, in tumor models, 192
Antibody candidates, for oncology application, 221
Anticancer antibodies, 183
Antitumor agents, 74, 78, 207
AP-2α protein
  in cancer formation, 87
  downregulation, by siRNA, 89
  microarray analysis, 92
  to p21 and DCBLD2 promoters, binding of, 92
  in tumor epithelial cells, 88
AP-2α siRNA, 91, 93
APL, see Acute promyelocytic leukemia
Apocrine carcinomas, 151, 167
Apoptosis, 89, 142
  inducers, 138
  induction of, 160, 162
  ROS and aging, P53/P66 pathway in regulation of, 178
ArrayScan™ system, 140
ARS, see Autonomously replicating sequences
Ataxia telangiectasia mutated (ATM), 62
Ataxia telangiectasia Rad3-related (ATR) checkpoint kinase, 62, 232
ATM/ATR-associated signaling machinery, 48

Subject Index

ATM/ATR checkpoint kinases
  biochemistry of, 48
  in DNA damage response, 49
ATM kinase activity,
  autophosphorylation of, 48
AT-rich tract, 3
Aurora kinase inhibitors, 138
Autonomously replicating sequences, 9
5 Aza 2′deoxycytidine, 239

B

Bacterial artificial chromosomes (BACs), 216
B cell neoplasms, 131
Bcr–Abl, nuclear accumulation of, 162
Bcr–Abl-induced lymphoblastic leukemia, 107–112
Bcr–Abl-induced myeloid, 108
BE2 cells, differentiation of, 181
Benign prostatic hyperplasia (BPH), 206
17Beta-hydroxysteroid dehydrogenases (17beta-HSD), 185
Beta-tubulin, 242
Biomarker clustering patterns, 146
Biotinylated retrovirus, 182
Biphosphonate zoledronic acid, 79
BJ-EHT cells, 41–42
BJ-ET cells, 21–22, 28, 30, 34
Bone marrow (BM) cells, 110, 231
Bone marrow progenitor cells, mice transplanted with, 233
Bone marrow transplantation (BMT), 110
53BP1, see p53 Binding protein 1
BRAF mutations, 203
BrdU incorporation in cell cultures, 142
Breast apocrine carcinomas, proteome expression profiling of, 151, 167
Breast cancers
  metastases of, 241
  and ovarian cancer suppression, 175
Breast cancer susceptibility gene product (BRCA1)
  for ATM and ATR dependent phosphorylation, 51
  defects in, 51
  in DNA repair, 50
  functional analysis of, 175
Bromodomain and ATPase Domain 1 (BRAD1)
  with chromatin and nuclear matrix, 63
  with MYC oncogene, 63
Bromodomain-containing protein BRD7, 210, 243
Brostallicin, 252

C

Caenorhabditis elegans, 229
Cancer
  epigenetics, transcription and signaling in, 59
  genome, structural abnormalities in, 150, 166
  and inflammation, 67, 103, 124
  mitotic progression in, 150, 166
  models of, 150, 166
  prevention, 102
  treatment, problem in, 101
  vaccines, MUC1 and CB1, 174
Carcinoembryonic antigen (CEA) protein, 204
Carcinogenesis, 101, 206
Caspase-3, 142
Cell culture and antibodies, 41
Cell-cycle inhibitors, 137
  hierarchical analysis of, 145
  mechanism of action of, 146
Cell-cycle markers, 138
Cell-cycle perturbations, 141
Cell cycle regulation, transcriptional mechanism of, 215
Cell division cycle protein 6 (Cdc6), 5, 8
Cell motility, 187
Cell signaling pathways, 129
Cellular markers, dose–response curves of, 144
Cellular senescence, 190
Centrosome-duplication inhibitors, 139
c-fos, in skin tumor formation, 117
C3G suppression mechanism, 249
Chemically-induced hepatocellular carcinoma (HCC), 125
Chemokines, 67, 103
  biological activities of, 253
  in human ovarian ascites fluid, 76
  in monocytes recruitment, 69
  receptor-expressing cells, 64
  as therapeutic target, 77
  at tumor site, 76

Chemoresistance, of tumor cells, 237
Chemotherapeutic agents, 58
Che-1 phosphorylation, 186
Chimpanzee serotype-based adenoviral vector, 204
Chk1 and Chk2 protein kinases, in G2 checkpoint, 199
Cholangiocarcinoma, 201
Chromatin binding, 5–6
Chromatin immunoprecipitation (ChIP), 92, 193, 235
Chromatin remodeling, at cis-regulatory elements, 196
Chromosomal V(D)J recombination, chromatin mediated regulation of, 236
Chronic myeloid leukemia (CML), 107
  Bcr–Abl in, 248
  P87C3G expression in, 248
  treatment, 162
Cisplatin, 235
Citron kinase (CRIK), 214
Colitis-associated cancer (CAC), mouse model of, 124
Colony stimulating factor-1 (CSF-1) in mammary epithelium, 77
Colorectal cancer
  metastatic and nonmetastatic, 224
  progression of, genetic and epigenetic changes in, 224
Coomassie Blue®, 54
Correlative microscopy, 169, 213
CpG island promoters, de novo methylation of, 121
CX3CR1-positive tumor cells, functional activity of, 195
Cyclin B1, 141, 145, 254
Cyclin-dependent kinase 1 (Cdk1)/cyclin B1 complex, 254
Cyclin-dependent protein kinases (Cdks), 3
Cyclooxygenase (COX), 79
Cytoplasmic c-Abl, 156

D

DAC, see 5 Aza 2′deoxycytidine
Dasatinib, 108
Dendritic cells (DC), 72
Deoxycytidine deaminase, 242
Deoxycytidine kinase, 242
Diethyl nitrosamine (DEN)-induced carcinogenesis, 125
Diethylnitrosamine-phenobarbital (DEN-Pb), 212
DNA
  binding proteins, ARID family of, 219
  chromosomal, 5
  encoded genetic information, 196
  methylation, 121
  recombination activity, 236
  sequencing system, 152, 168
  synthesis, 4–5, 9
DNA damage
  checkpoint pathway, 47
  induced apoptosis, 162
DNA damage response (DDR)
  53BP1 in, 51
  BRCA1 in, 50
  c-Abl in, 157
  MDC1 in, 49
  oncogene activation, 118
  proteins activities during, 48
  regulation of, 49
  in senescent human fibroblasts, 190
  Tip60 in, 184
DNA-damaging agents, 155
DNA double-strand breaks (DSBs), 62
DNA ligase I, 4
DNA methyl-transferases, 239
DNA polymerase α/primase, 4
DNA replication
  in mammalian cells, 232
  of Simian Virus 40, 3–4
  in Xenopus laevis, 4–7
DNA topoisomerase inhibitor SN-38, 145
Double strand breaks (DSB), 52
Doxycycline-inducible ALK–sh RNA, 131
Drosophila germ cells, 39
Drosophila histone deacetylates (DHDAC1)
  deacetylase activity of, 189
  silencing of, 189
Drosophila melanogaster, 229, 246
Drug discovery process, 137
Drug induced cytostasis, 112

E

Early palindrome (EP), 3
E-cadherin, 197, 201
Eidermal stem cells, mobilization of, 115
Embryonal carcinoma (EC), 29
Embryonic stem cells
  histone methylation in, 216
  miRNA372-373 in, 39–40
Endo-5′-nucleotidase (5′-NT), 242
Epidermal growth factor (EGF), 75
Epidermal growth factor receptor (EGFR)
  immunohistochemical method of, 217
  in malignant pleural mesothelioma (MPM), 217
  signal transducers, 201
Epigenetic silencers, 121, 200
Epigenetic therapy, in cancertreatment, 121
Epithelial cancers, models of, 150, 166
Epithelial-mesenchymal transition (EMT), 60, 104
Epithelial ovarian carcinomas (EOCs), 197
Epithelial thyroid tumors, chemokines treatment of, 253
Escherichia coli, 7, 216
Estradiol (E2), in normal breast and tumor tissues, 185
Estrogen receptor modulator, 188
Estrogen sulfates (ES), 185
Etoposide, 232
Ets transcription factors, PEA3 group of, 177
Eucaryotic chromosome, 5, 218
Eukaryotic genomes
  replication of, 149
  transcription factors (TFs), 193
Eukaryotic protein complex for human cell proliferation, 198
Eukaryotic replicator sequence, 9
Eu-myc transgenic mice, B-cells of, 118
Extra-cellularly regulated kinases (ERK), 97–98, 249

F

Flap endonuclease 1 (Fen1), 4
Flow cytometry, 137
Fluorescence intensity histograms, cellular marker, 142
Fluorescence light microscopy (FLM)
  limitations of, 169, 213
  multi section 3D reconstruction of, 169, 213
Fluorescence staining intensity, 140
Fluorescent dyes, 138
Fra-1 and Fra-2 (Fos-related antigen 1 and 2), as p53 modulators, 238
Fractalkine receptor, 195
Fumarate hydratase (FH), tumor suppressor gene, 225, 251

G

Geminin, 5
Gene abnormalities, 224
Gene expression, ubiquitin-like proteins in, 122
Gene expression profile (GEP), 133
Gene function, retrovial array to study, 182
Gene regulatory factors, 196
Genetic lesions, 59
Genome Sequencer 20 System, 152, 168
Genome stability, 218
Genome-wide chromatin-based isolation, of active cis-regulatory DNA elements, 196
Genomic DNA, 41
Genomic repression, regulation of, 123
Germline heterozygous mutations, 225, 255
Germline inactivating mutations, 251
Gliobastoma multiforme (GBM), telomerase involvement in, 209
Glucose-regulated protein 94 (Grp94) stress promoter, 231
Glutathione-*S*-transferase P1 (GST) promoter, 252
Glycine-arginine rich (GAR), 51
Green fluorescent protein (GFP), 110
Grf1-deficient mice, pancreatic islets of, 250
Guanine nucleotide exchange factor (GEF) C3G, 249–250

H

HDACs, *see* Histone deacetylases
Heart ischemia-reperfusion injury, 205
HeLa cells, 88, 91
Hematopoietic lineages, miRNAs in, 191

Hematopoietic neoplasms, 109
Hematopoietic oncoproteins, p53
  functional inactivation by, 227
Hematopoietic stem cells, 108
Hepatocyte growth factor (HGF), 179
Hereditary leiomyomatosis and renal-
  cell carcinoma (HLRCC), 225
Hierarchical cluster analysis, of 20 cell-
  cycle inhibitors profiles, 145
High-content analysis (HCA)
  application of, 138
  for automated analysis of subcellular
    events, 137
  multiparameter approach, for cell-
    cycle analysis, 138
High-content profiling (HCP), 139
High-content screening (HCS), 138
Histone acetylation, 7, 215
Histone acetyltransferases (HATs),
  184, 207
Histone deacetylases, 207, 230
Histone deacetylases 1
  in human tumor cell proliferation, 230
  sumoylation on, 247
Histone deacetylases 1/2 interacting
  protein MTA2, 207
Histone H3 phosphorylation, 141
Histone methyltransferases (HMTs), 200
HSCs, see Hematopoietic stem cells
hSNM1B (telomeric protein), with
  TRF2, 218
Human epithelial cells, 101
Human prostate cancer (PCa), 97, 194
Human replication origins, 149
Human tumor antigens, 174

I

IgG protein, 221
IκB kinase β (IKKβ) deletion, in
  myeloid cells, 124
Imatinib, 108, 111
Imatinib-treated mice, 112
Immunocompetent mice, Arf
  inactivation, 111
Immunosuppressive cytokines, 76
Inflammation-induced tumor
  progression, 124
Inflammatory cytokines and cancer,
  64, 69
ING1 protein, 222

INK4A/ARF tumor suppressor
  locus, organization and
  regulation of, 107
Interleukin (IL), 69, 109, 231
Intermolecular disulfide bridges, 52, 54
Intracellular proteins, proteolysis of, 208
Invasive ductal carcinoma, 185

J

Janus (JAK) kinases, 112
Junctional adhesion molecule-A (JAM-
  A), in controlling PMN
  diapedesis, 205
c-Jun N-terminal kinase (JNK)
  phosphorylation, 160

K

Keratinocytes, 214
Kinetochores, 220
Kupffer cells, 125

L

LArge Tumor Suppressor homolog 2
  (LATS2)
  by miR-372 and miR-373, inhibition
    of, 36
  suppression of, 34
  YFP-competition assay of, 35
Lesions, early detection of, 101
Lethal lymphoid leukemia, 110
Leukemias
  in recipient animals, 111
  in vivo epigenetic pharmacotherapy,
    239
Linomide, 78
Liver cancer development, AP-1
  (c-Jun/c-Fos) in, 212
Liver tumors, p53 in, 212
LNCaP cells, with brostallicin, 252
Lung adenocarcinomas, 116
Lung cancers, 234
  biological features of, 242
  diagnostic and prognostic markers of,
    116
  miRNA expression profiles for, 116
Lymphoid neoplasia, in transgenic mice,
  131
Lymphoid-specific proteins, 236
Lymphomagenesis, Notch suppresses
  p53 in, 255

M

Macrophage colony stimulating factor (M-CSF), 72
Macrophages, forms of, 69
Mad2 protein, activation in SAC, 220
Malignant melanoma, molecular signature of, 203
Malignant pleural mesothelioma, 217
Mannose receptor (MR), 74
MAPK, see Mitogen activated protein kinase
Matrix metalloproteinase-3, 60
MCF7 cells, 88, 91
Mechanism of action (MOA), compounds with similar, 146
Mediator of DNA damage checkpoint protein 1 (MDC1), 49–50
MEF, see Mouse embryonal fibroblasts
Meiosis, role of p63 in, 240
β-Mercaptoethanol, 54
Messenger ribonucleoprotein (mRNP) biogenesis, 198
MET, in human tumor formation, 251
Metastatic breast tumor model, 241
Metastatic tumors, 224
Methylated lysine 79 of histone H3 (H3-K79me), 52
Met proto-oncogene, in human tumors and in metastasis, 179
Microarray-based gene expression signatures, 130
Microphthalmia-associated transcription factor, 59
microRNAs, see miRNAs
Microtubule inhibitors, 140
Mid-blastula transition (MBT), 7
Minichromosome maintenance proteins (Mcm), 5–6
miR-93 and miR-302a-e, 38
miR-372 and miR-373 expression
  CDK2 activity in TGCT, 30
  in cellular transformation, 39
  in embryonic stem cells, 39–40
  in human cancer, 29–33
  with oncogenic RAS, 35
  on p53 activation, 27
  protects from oncogenic stress, 22
  with RASV12, 27, 38
  regulate LATS2 expression, 34–36
  in TGCT development, 39
  transforms primary human cells, 26–29
miR-371–373 genomic organization, 25
miRNA expression library (miR-Lib) sensitivity of screens with, 19
miRNA 372-negative seminomas, 33
miRNAs
  biogenesis of, 123
  detection, 41
  functional genetic screens for, 21, 35
  for lung cancers, 116
  mechanism of action, 18
  minigenes, 40
  in myeloid differentiation, 191
  retroviral vector for, 19
  RNase protection assay (RPA), 19
miR-Vec system
  functionality of, 19
  inhibits oncogene-induced senescence, 23
Mitochondrial tumor suppressor, 225
Mitogen activated protein kinase, 97
Mitotic chromosome condensation, 142
MMP-3, see Matrix metalloproteinase-3
Monocyte macrophage lineage, 68
Mononuclear phagocytes, 68
Monoubiquitination, 173
Monouniquitin, 208
Mouse bone marrow transplantation model, 110
Mouse embryonal fibroblasts, 134
MPM, see Malignant pleural mesothelioma
mRNA-expression array analysis, 34
mTOR-containing proteins complexes, regulates S6K and Akt PKB, 61
Multiparametric cellular analysis, 137
Multiple myeloma (MM), 188
Mutant p53 proteins, 215, 237
Myc binding, epigenetic determinants of, 193
Myc oncoprotein, 202
c-Myc physiological function, in adult mice tissues, 228
Myeloid differentiation, microRNAs in, 191
Myeloma-induced bone lesions, 188

N

Nanostructured TiO$_2$ retroviral array, 182
Neoplastic tissue, 64, 103
Neuroblastoma
    cell lines, 181
    PDGF receptor beta (PDGFRB) in, 235
Neuroendocrine molecules, 96
Neuronal differentiation, 181
Neuropeptide Y,
    antiproliferative effects of, 97
    on cell growth, 98
    immunopositivity for, 97
    in prostate cancer, 96–97
Neuropeptide Y–Y1 receptor system, on prostate cancer cell growth, 194
Neutrophil chemotactic proteins, 72
NF-κB
    activity, in myeloma cells, 188
    transcription factors, activation of, 124
Nicotinamide adenine dinucleotide, biosynthesis of, 79
NLS, *see* Nuclear localization signals
Nonclathrin pathway, 173
Non-Hodgkin lymphomas (NHLs), 130
Nonseminomas, 29
    miR-372 expression detection in, 32
    in situ hybridization on, 32
Nonsmall Cell Lung Cancer (NSCLS), 78, 242
Notch-induced neoplastic disease, 255
NPM-ALK
    cellular transformation, 131
    down-modulation of, 131–132
    gene expression signature, 133
    p53 functional down-regulation by, 227
    phosphorylation of, 131
    structure and function of, 227
    transforming properties of, 134
NPM gene, 130
NPY, *see* Neuropeptide Y
NSCLC, *see* Nonsmall cell lung cancer
Nuclear area perturbation, 145
Nuclear localization signals, 159
Nuclear morphology analysis, 138, 141
Nuclear proteins, activation of, 162
Nuclear translocation, 157, 162
Nucleoprotein complexes, 218
Nucleus/chromatin shape, 138

O

Okazaki fragment maturation, 4
Oligomerization, of 53BP1, 52
Oligonucleotide glass arrays, 206
Oncogene-activated p53 pathway, 38
Oncogene-induced senescence (OIS), 2, 219
Oncogenes, 18, 219
    activation of, 223
    fusion proteins, 176
    H-RASV12, 18
    pathways analysis, gene expression profiles for, 130
    in tumorigenesis, 19
    Tyr-kinase fusion proteins, 227
Oncology, application in, 221
Oocytes, DNA damage in, 240
Organ homeostasis, 115
Origin recognition complex (ORC), 5–6
Ovarian surface epithelium (OSE), 197

P

Pancreatic islets, in GRF1-deficient mice
    differential gene expression profile of, 250
    Ras activation in, 250
Pancreatic tumor cell adhesion, 195
Papillary renal cell carcinomas, 251
p53 binding protein 1
    carboxyl terminus of, 51
    complex formation of, 54
    oligomerization, 52
    orthologs of, 52
    purification of, 54
    Western Blot method for, 53
    whole cell lysates for, 53
PCNA, *see* Proliferating cell nuclear antigen
PDGF receptor beta (PDGFRB)
    activated by GATA, 181
    transcriptional regulation of, 235
PEA3 group members, regulation of, 177
Pentanucleotides, 3
p53 family proteins, transcriptional activity of, 210, 243
Pharmacogenomics, 242
Pharmacologically active compounds, MOA profiling of, 139

Philadelphia acute lymphoblastic leukemias (Ph+ ALLs), 110
Philadelphia chromosome (Ph), 108
Phosphoinositide 3-kinase related kinases (PIKKs), 48–49, 52
Phosphorylated extracellular kinase (ERK)1/2, 194
Phospho-S6 (Ser 235/236), 142
PML-RARα fusion protein, to leukemogenic process, 180
p53/NF-Y protein complex, 215
Podoplanin, in breast cancer cells, 104
Polarized macrophages, 68
Polo-like kinase 1 (Plk1), 254
Polycomb-mediated transcriptional silencing, RNA interference in, 246
Polymorphonuclear leukocytes (PMN), 205
Polyomavirus, 10; see also *Simian Virus 40*
p53 protein
  antagonists of, 238
  in genomic stability and cellular homeostasis, 210, 243
p53–p66Shc signaling pathway, 178
pRB/E2F pathway, 63
PRDM16 protein, 233
Pre-B cells, 109
  Arf-p53 checkpoint in, 110
  intra-peritoneal (IP) injection of, 111
Preleukemic bone marrow, 180
Pre-malignant lesions, 101
Proliferating cell nuclear antigen, 4
Prostate cancer (PCa), 96
  gene expression profiling on, 206
  multiple ETS family transcription factors in, 206
  treatment of, 252
Protein complex mutant p53/p73, in tumor cells, 237
Protein phosphorylation, 158
Protein–protein interactions
  regulation of, 158
  SUMO in, 247
14-3-3 proteins, 158–160
p66shc protein, redox property of, 178
pSUPER cells, 91

R
Rad9
  for DNA damage signalling, 49
  forms of, 52
  MDC1, BRCA1, and 53BP1, 49
  phosphorylation, 49
Raloxifene-induced myeloma cell apoptosis, 188
RaLP expression
  down-regulation of, 244
  in melanoma progression, 244
Rapamycin, 61
RASV12-induced growth, 22–23
Reactive oxygen intermediates, 74
Reactive oxygen species, 60
Receptor tyrosine kinases, 173
Recombinant scFv antibody, 183
Recombination signal sequences, 236
Replication factor C (RF-C), 4
Replication protein A (RPA), 4
Replicative proteins, 232
Retinoblastoma tumor suppressor (pRb), 229
Retroviral microarrays
  for drug target identification, 182
  for gene therapy application, 182
Retrovirus, 41
Rho signaling
  in cell-cell adhesion, 214
  in keratinocyte differentiation, 214
Ribonucleotide reductase, 242
Ribosomal protein S6, 142
RNAi libraries, 150, 166
RNAi-mediated protein knockdown, 230
RNA-induced silencing complex (RISC), 123
RNA interference technology, 18, 130
ROIs, *see* Reactive oxygen intermediates
ROS, *see* Reactive oxygen species
Roscovitine-induced apoptosis, 245
RSS, *see* Recombination signal sequences
RTKs, *see* Receptor tyrosine kinases

S
Saccharomyces cerevisiae, 9, 49, 149
Scavenger receptor-A, 74
Secreted protein acidic and rich in cysteine, 79

Senescence associated DNA damage foci (SDFs), 190
Senescence associated (SA-β-Galactosidase), 22
Serine/threonine kinase, 211
  Cdc7, 7
  residues, phosphorylation level of, 159
SET domain-negative form, 233
Simian Virus 40
  and disease, 10–11
  DNA replication of, 3–4
    advantages and limitations, 9
    initiation of, 8
  genetic information of, 3
Smokers, sputum of, molecular alterations in, 234
Soft agar assay, 42
SPARC, *see* Secreted protein acidic and rich in cysteine
Spindle assembly checkpoint (SAC), 184, 220
SpotFire software, 145
sPRDM16, in murine bone marrow progenitor cells, 233
SR-A, *see* Scavenger receptor-A
Stability diverse transcriptional complexes, SUMO in, 247
Staurosporine-induced apoptosis, 90
Staurosporine (STS), 89
Stem cell epigenetics, 216
Streptavidin-biotinylated viral vectors, 182
Stromelysin-1/matrix metalloproteinase-3 (MMP-3), 60
Sulfolobus solfataricus, 3
Sulfotransferase, 185
SUMO protease SENP1, structural analysis of, 122
Sumoylation, 177, 247
SV40, see Simian Virus 40

T

TAM, *see* Tumor-associated macrophages
Tamoxifen, 26, 188
T-antigen, 3, 8, 18
TAR, *see* Transcription associated recombination
T cell activation, suppressor of, 79

T-cell lymphoma, initiation of, 255
T cell receptors, 236
TDFC, *see* Tumor-derived chemotactic factor
Telomeres
  activity, in primary human endothelial cells, 209
  binding protein TRF2, 115
  catalytic component of, 115
  DNA binding factors, 218
  length, 115
Testicular germ cell tumor (TGCT), 19
  miR-371–373 cluster detection in, 29
  p21-mediated cell cycle arrest in, 33
  p53 status in, 32–33
  types, 29
Tetracycline-inducible Chk1 siRNA clones, 199
Tetracycline inducible mouse model, 255
Tetracycline–responsive Omomyc transgene, 228
Thymidine phosporylase (TP), 75
Tip60 heterozygosity, 118
Tissue homeostasis, 216
Tissue specific (TS) stem cells, in histone methylation, 216
TNF-α, *see* Tumor necrosis factor-α
Topoisomerase I (topo I), 4
Transcription associated recombination, 198
Transforming growth factor-β (TGF-β) signaling, 211
T regulatory cells (Treg), 75
TRF2 mice, skin phenotypes, 115
Trithorax (trx) HMT, 200
Tudor domains, 52
Tumoral phenotype, 130
Tumor-associated macrophages
  adaptive immunity by, 75
  infiltration, 71
  NF-κB activation, 74, 77
  polarization, 68
  protumoral role of, 72–75
  stimulatory effects of, 78
  suppressive mediators produced by, 75
  as therapeutic target, 76–79
  in tumor hypoxia, 78
  VEGF-C production by, 74

Tumor cells
  cytoskeleton of, 241
  growth and survival of, 134
  invasion, 104
  vs. normal cells, 174
Tumor-derived chemotactic factor, 69–71
Tumor growth, 68
Tumor hypoxia, 78
Tumorigenesis, 26
  in *Drosophila* eye, 200
  miR-372 and miR-373 in, 29–33, 38
  RASV12 in, 38
Tumorigenicity, 19, 42
Tumor-infiltrating B lymphocytes, oligoclonality of, 183
Tumor microenvironment, 60, 231
Tumor necrosis factor-α, 64
Tumor-specific antibodies, 183
Tumor-suppressive mechanisms, 245
Tumor suppressor genes, 58
  inactivation of, 223
  in p16INK4a/pRB and p53 pathways, 223
  p53 protein, 210, 243
Tumor surface antigens, 183
Tyrosine kinase inhibitor imatinib, 111
Tyrosine kinase inhibitors sensitivity (TKIs), 201

U

Ubiquitination, 177
  of EFGR, 173
  functions, 208
Ubiquitin-like modifier SUMO, 122
U2-OS cells
  extract from, 54
  proliferation, 230
  representative figures of, 142–143

V

Valproic acid (VPA), 239
Variant hepatocyte nuclear factor 1 (vHNF1), 197
Vascular endothelial cells (ECs), 211
Vascular endothelial growth factor (VEGF), 70, 72, 75
  production of, 74
  in TAM recruitment, 78
Vascular endothelial (VE)-cadherin, 211
Viral proteins, 3

X

Xenograft tumors, growth patterns of, 131
Xenopus laevis, 3
  chromatin binding of, 5
  DNA replication of, 4–7
    advantages and limitations, 10
    initiation of, 8
  eggs, 10
  genetic information, 4
  heterohexamer in, 7
Xeroderma pigmentosum (XP) syndrome, 115

Y

Yeast, 47, 49
  53BP1 in, 51
  budding and fission, 52
  chromosomal origin activity in, 149
Yondelis (Trabectedin), 78

Z

Zebularine, 252
Zygotic genes, 7